Fetal

Development

and

Cannabis

Fetal Development and Cannabis

Authors

Abdulrahman Aldada

Annabella Stoll-Dansereau

Anoushka Kaliyambath

Eline El-Awad Gonzalez

Erwin Kwok

Holly Steen

Noah Serethe

GM
PRESS

First Printing: 2023

Typeset and cover design by Clare Dalton

Print ISBN: 978-1-77889-059-8

eBook ISBN: 978-1-77889-060-4

Golden Meteorite Press

103 11919 82 St NW

Edmonton, AB T5B 2W3

www.goldenmeteoritepress.com

Contents

Foreword

Annabella Stoll-Dansereau

Fetal Alcohol Syndrome is a widely understood and recognised condition that arises when a pregnant woman consumes alcohol while in the womb and a very similar phenomenon occurs with cannabis called Fetal Cannabis Syndrome. Through this book, we will highlight these similarities and educate the reader on the actual interactions this drug has on fetal brain development touching on the effects on the nervous system and the whole body. We will also look at how the prevalence of this condition varies between population demographics and if it is changed through the legalisation of the drug or during COVID. Additionally, we will compare outcomes when cannabis is used in large versus smaller doses and touch on prevention methods to reduce the use of cannabis by pregnant individuals. We hope the reader will come away with a better understanding of this syndrome and help build awareness surrounding this topic.

Introduction to Fetal Development and Cannabis

Abdulrahman Aldada

The Increasing Legalisation of Cannabis and the Potential Effects on Fetal Development

Introduction

Cannabis, usually known as marijuana, is a psychoactive plant substance used medically and recreationally. It is made from the cannabis plant's flowers, leaves, and stems and includes tetrahydrocannabinol (THC). Depending on the strain, the amount of THC in cannabis might vary, but it is the primary psychoactive ingredient (Bridgeman & Abazia, 2017). For thousands of years, individuals have used cannabis for therapeutic, spiritual, and recreational purposes. Ancient Chinese physicians employed the herb to cure several conditions, including nausea and pain alleviation. Cannabis usage and possession became unlawful in the United States in 1937 due to the *Marijuana Tax Act* (History.com Editors, 2017). Cannabis legalisation has become increasingly popular in Canada recently. While some provinces have legalised the custody and usage of marijuana, others have authorised it for recreational use. In addition, several provinces have permitted the usage of medical cannabis for a variety of ailments.

Fetal Cannabis Syndrome has prompted worries about cannabis' possible effects on fetal development as its legalisation spreads. Exposure to cannabis during pregnancy can cause Fetal Cannabis Syndrome, a disorder marked by cognitive decline, developmental delays, and other health problems. However, it is known that THC can penetrate the placental barrier and enter the fetal circulation, potentially altering fetal development. Research on the consequences of prenatal cannabis consumption is relatively restricted. The public and medical professionals should be aware of the possible dangers of fetal cannabis syndrome and prenatal cannabis exposure. Given the growing tendency toward legalising cannabis, this is particularly crucial. It is essential that healthcare professionals inform expectant mothers about the possible hazards of cannabis use and that public health authorities create plans to address Fetal Cannabis Syndrome and lessen its prevalence. Fetal Cannabis Syndrome is a growing concern, so it is essential for medical professionals and the general public to be informed of the expanding legalisation of cannabis and its potential consequences on fetal development. More research is required to comprehend the severity of this problem and create plans to lessen its occurrence (et al., 2017).

Overview of Fetal Cannabis Syndrome

A medical disease known as Fetal Cannabis Syndrome is thought to result from prenatal cannabis exposure. It is characterised by various physical and behavioural issues, such as low birth weight, delays in cognitive development, and aberrant motor development. Although the precise origin of Fetal Cannabis Syndrome is unknown, there is mounting evidence that prenatal cannabis acquaintance can be a risk factor. The first description of Fetal Cannabis Syndrome in the medical literature appeared in 2017, making it a relatively recent medical ailment. Low birth weight, cognitive delays, motor abnormalities, and attention deficit hyperactivity disorder are among the physical and behavioural issues that the illness is characterised by and can impact a child's development (ADHD). Although the precise origin of Fetal Cannabis Syndrome is unknown, there is mounting evidence that prenatal cannabis acquaintance can be a risk factor. Fetal Cannabis Syndrome can result in low birth weight, inadequate physical development, and a small head circumference as physical signs. The most prevalent physical indication of Fetal Cannabis Syndrome is low birth weight, typically brought on by intrauterine growth restriction (IUGR). Inadequate nutrition or other causes can cause IUGR, a disease in which the fetus does not usually develop inside the uterus. Low birth weight is just one of the health issues that might result from this.

Regarding cognitive delays, Fetal Cannabis Syndrome can result in executive dysfunction, poor problem-solving skills, and language difficulties. Language delays, frequently linked to Fetal Cannabis Syndrome, can make it challenging to grasp and use. A lack of problem-solving skills might make it difficult to reason and think abstractly. Planning and arranging work can be difficult when executive functioning is poor. Fetal Cannabis Syndrome can result in decreased muscle tone, poor coordination, and poor balance in motor abnormalities. A lack of muscular tone might make it harder to control your emotions and keep your posture. Ineffective coordination can make completing tasks like writing and typing challenging. Standing and walking can be difficult as a result of poor balance. Fetal Cannabis Syndrome might result in focus issues, hyperactivity, and impulsivity regarding behavioural issues. Problems paying attention might make it difficult to concentrate on work and follow instructions. Instability and trouble staying motionless can result from hyperactivity. Impulsivity can make it difficult to regulate emotions and adhere to regulations. A child's physical and behavioural development may be negatively impacted by the significant medical illness of Fetal Cannabis Syndrome in the long run. Although the precise origin of Fetal Cannabis Syndrome is unknown, there is mounting evidence that prenatal cannabis acquaintance can be a risk factor. Expectant moms should discuss cannabis use with their healthcare physician and be aware of any hazards of using marijuana while pregnant.

Potential Effects of Cannabis Legalisation on Fetal Development

Cannabis use is rising in popularity as more countries have legalised it in recent years. There may be effects on the health and growth of a fetus from the legalisation of cannabis. With an emphasis on prenatal development, the results of cannabis usage on fetal development, and the potential formation of Fetal Cannabis Syndrome, this article will examine the possible impacts of cannabis validation on fetal growth. One of the essential effects of cannabis legalisation may be expanding pregnant women's access to cannabis. The risk of cannabis usage during pregnancy may rise due to pregnant women's increased propensity to consume the drug. The consequences of cannabis use on fetal growth during pregnancy remain unknown because it is a psychotropic drug. Cannabis usage during pregnancy has been linked to lower birth weight and a higher chance of premature delivery, according to studies. Cannabis may also be connected to a shorter gestational period and a higher event of stillbirth. We don't fully understand how cannabis usage affects embryonic development. Cannabis usage in gestation has been linked to advanced jeopardy of cognitive and behavioural issues in offspring, according to studies. A higher risk of deformities and mental and memory problems may also be linked to cannabis usage during pregnancy. A higher incidence of autistic spectrum conditions may also be related to cannabis use during pregnancy. Increased Fetal Cannabis Syndrome is a potential side effect of cannabis legalisation.

Using cannabis during pregnancy can cause various behavioural, cognitive, and physical issues that make up this syndrome. Low birth weight, premature delivery, and an elevated risk of behavioural, mental, and physical problems are all signs of fetal cannabis syndrome. Fetal Cannabis Syndrome may also be linked to memory and learning problems, as well as a higher chance of autistic spectrum condition. Cannabis legalisation may affect fetal development, yet this is still mostly unknown. However, more accessible access to cannabis for expectant mothers could increase the incidence of Fetal Cannabis Syndrome. It is also critical to remember that further research is required because the effects of cannabis usage on prenatal development are currently poorly understood. Finally, it is crucial to take into account how legalising cannabis can affect fetal growth and to make sure that expectant mothers are aware of any possible dangers from using the drug during their pregnancies.

Arguments for and Against Cannabis Legalisation

A. *Arguments for Legalising Cannabis:*

Cannabis legalisation has long been a hotly contested issue, with supporters from both sides. Proponents of cannabis legalisation point to several potential benefits, including increased tax revenue, improved public health, and reduced crime rates. Here, we will discuss some main reasons why marijuana ought to be legal. The legalisation of cannabis presents an excellent opportunity for governments to generate additional tax revenue. By legalising and taxing cannabis, governments will benefit from the considerable profits associated with the sale of marijuana. This could provide a much-needed income source, which can be reinvested into the community in various ways. Legalising cannabis could also help to improve public health. By making marijuana legal, it can be regulated, thereby ensuring that only tested, safe products are available on the market. This could help reduce the health perils related to marijuana, like the potential for respiratory problems. Legalising cannabis could also lead to a reduction in crime. By regulating the sale of marijuana, it would become much more difficult for criminals to obtain and sell the drug. This could decrease drug-related crime, allowing the police and other law-implementing authorities to focus their efforts on extra solemn crimes.

1. Boosted Tax Collection.

The possibility for more tax income is one of the most well-liked justifications for legalising cannabis. Cannabis can bring in much-needed tax income for the government when it is controlled and taxed similarly to alcohol. For instance, after legalisation in 2018, Canada has generated over $1 billion in cannabis-related taxes. Again, since it was legalised in Colorado in the United States in 2012, cannabis taxes there have brought in over $1 billion. Programs for public health, education, and other public services can be paid for with this money (Krishna, 2022).

2. Enhancement of Public Health.

The health of the entire population could improve if cannabis were legalised. Cannabis users will be able to receive a product of more excellent quality after the drug is subject to enhanced regulation and control, both of which are made possible when cannabis is legalised. The social stigma associated with cannabis use is reduced when it is legalised, which may make it informal for people who use marijuana to search for aid if they require it. By decriminalising cannabis, the burden placed on the criminal justice system could be reduced, resulting in the release of resources that could be put toward the prosecution of more serious crimes (Farley & Orchowsky, 2019).

3. A Decline in Crime.

Several possible benefits could result from legalising cannabis, including a crime reduction. When cannabis, for example, is made legal, it is removed from the hands of those who sell it on the black market. It follows that those who are now engaged in the unlawful distribution of cannabis will no longer be able to profit from their activities. Consequently, associated crimes such as robberies and assaults may decrease. Legalisation may also result in a drop in the figure of individuals arrested for possessing illegal substances (Stohr et al., 2020).

B. Arguments Against Legalising Cannabis:

Several possible benefits come with legalising cannabis, but there are also a few negatives to consider. Opponents of legalising cannabis raise concerns about the potential risks to public health and safety and problems that could arise economically. Let's take a look at some of the arguments against making marijuana available for recreational use (SAMHSA, 2019).

1. Risks to public health and safety

One of the most potent arguments against making cannabis consumption legal is the risk that it could pose to public health and safety. Cannabis use has been related to an amplified peril of addiction and mental health problems, as well as an increased risk of being involved in a car accident or suffering from another type of injury. The increased availability of marijuana to young people due to the legalisation of cannabis could also make the drug more hazardous to their health and safety (Wilkinson, 2013).

2. Financial Issues

Another point that works against cannabis legalisation is the possibility that it may harm the economy. For instance, there is an indication to suggest that marijuana use can lead to a decrease in productivity and increase the number of days missed from work. Furthermore, legalisation can lead to a rise in the sum of individuals who use the substance, which would drive up the expense of healthcare for businesses (Popovici & French, 2013).

3. Social Implications

The decriminalisation of cannabis comes hand in hand with a host of social issues. Opponents of legalisation point to the possibility that it may increase cannabis consumption, which may, in turn, make societal problems like substance abuse and homelessness even more severe. In addition, it may cause the drug to become more acceptable in society, increasing the number of people who use it (Svrakic et al., 2012).

There are a variety of complex arguments that may be made in backing or obstructing the legalisation of marijuana. Those who favour legalisation point to potential benefits like increased tax receipts, improvements in public health, and a reduction in criminal activity. On the other hand, those who are against legalisation argue that it could lead to issues in both the economy and society, in addition to potential risks to the public's health and safety. In the end, the problem regarding the legalisation of cannabis is a difficult one that needs to be carefully evaluated and taken into account (Stohr et al., 2020).

Literature Review

1. *Studies on the Impact of Marijuana Use on Gestation*

Recent study efforts have been concentrated on finding out what the consequences of using cannabis while pregnant are. Numerous studies have been conducted to determine the levels of cannabis consumption that are harmful to an unborn kid and a developing fetus—according to the findings of this study, consuming cannabis. In contrast, pregnancy can have several immediate and lasting impacts on the unborn child. The impacts of cannabis consumption in gestation on the development of the fetus were the subject of research shown in the United States in 2018. The work finds that prenatal marijuana use was related to smaller head circumferences and slower growth in pregnancy's first and second trimesters. The research also found that cannabis use during pregnancy had a more negative impact on male offspring's physical and mental development than female offspring.

Additional research has been conducted to explore the impact that marijuana has on the growth of the fetus. In 2019, a study was born in the United Kingdom to determine how prenatal experience with cannabis affects the growth of the nervous system. According to the study's findings, using cannabis during pregnancy is associated with having a smaller brain and less connection in the brain's white matter. The research also found that executive functioning and attention impairments were associated with this diminished connection (Grant et al., 2018). A more recent inquiry into the impact of cannabis usage during gestation on the physical health of newborns was conducted in the Netherlands in 2020. This study was published in the year 2020. According to the findings of the study, using marijuana in gestation was related to an amplified peril of early birth as well as lower birth weight. The research also found that using cannabis in gestation was related to an amplified risk of gastrointestinal and respiratory issues in the offspring.

2. *Studies on the Prevalence of Fetal Cannabis Syndrome*

Research has also been done to establish the prevalence of the disorder known as Fetal Cannabis Syndrome, which may be caused by the use of cannabis by the mother while she is pregnant. This is because cannabis usage during pregnancy may negatively affect the baby. Cannabis usage during pregnancy is linked to a variety of health problems for both the mother and the developing baby. Numerous investigations have been carried out to obtain a deeper comprehension of the frequency of occurrence of Fetal Cannabis Syndrome. A study was conducted in Canada in 2018 to establish the prevalence of Fetal Cannabis Syndrome in a group of children ranging in age from five to nine years old. The participants in the study were all children (Allison Bradbury, 2019). The outcomes of the work proved that children whose mothers had used cannabis while they were pregnant had a significantly higher risk of developing FAS than other children did. According to the study's findings, the severity of Fetal Cannabis Syndrome symptoms was shown to be amplified when cannabis was ingested during pregnancy. This was one of the key takeaways from the research.

In the United States of America in the year 2020, recent studies have been passed out to estimate the prevalence of Fetal Cannabis Syndrome in a populace of pregnant women in the country at the time (Anderson, 2014). The study showed pregnant women who used cannabis had a much higher risk of having a child affected by Fetal Cannabis Syndrome than other pregnant women. According to the study's findings, the severity of Fetal Cannabis Syndrome symptoms was shown to be amplified when cannabis was ingested during pregnancy. This was one of the key takeaways from the research. Studies on the impact of using marijuana while pregnant and the prevalence of Fetal Cannabis Syndrome have shown, in general, that there are several short-term and long-term repercussions associated with this practice. These studies have focused on the impact of using marijuana in gestation. These studies have been carried out on pregnant women and the unborn children that those women carry. Additionally, studies have revealed that a woman's likelihood of having a child with fetal alcohol spectrum disease is increased when she uses cannabis while she is pregnant (Fetal Cannabis Syndrome). As a consequence, it is of the utmost significance that pregnant women be made aware of the potential hazards of using cannabis. At the same time, they are pregnant and refrain from using cannabis during their pregnancies (Roncero et al., 2020).

Conclusion

In conclusion, the increasing legality of cannabis has substantial repercussions for the development of the fetus, and these possible repercussions should not be taken lightly. In addition, the legalisation of cannabis has enormous implications for the development of society as a whole. Even though there is just a limited amount of evidence to support fetal cannabis syndrome, there is a growing concern that it may become more widespread in the years to come. Because it is not yet understood how the drug will affect the developing fetus, additional research is required to determine the full degree of the dangers linked with smoking cannabis while pregnant. These dangers comprise but are not partial to the following: Up until that moment, pregnant women or breastfeeding should exercise great caution if they consider using cannabis since it may have detrimental implications on the baby who is still developing. If you are considering cannabis, you should exercise extreme caution (UNODC, 2018). As the number of states that have passed legislation to decriminalise cannabis use continues to rise, it is more important than ever for medical professionals to be aware of the Fetal Cannabis Syndrome and to educate their patients about the potential risks associated with using cannabis while pregnant. If a woman is pregnant or breastfeeding, it is of the utmost importance for her to be alert to the possible perils related to the use of cannabis and to have a conversation about the topic with her healthcare provider before initiating any activities that involve cannabis.

References

ACOG. (2021, August 21). *Marijuana Use During Pregnancy and Lactation.* Www.acog.org. https://www.acog.org/clinical/clinical-guidance/committee-opinion/articles/2017/10/marijuana-use-during-pregnancy-and-lactation

Allison Bradbury. (2019, March 27). *Marijuana and Pregnancy.* Samhsa.gov. https://www.samhsa.gov/marijuana/marijuana-pregnancy

and, E., Health, Board, & Evidence, A. (2017, January 12). *Prenatal, Perinatal, and Neonatal Exposure to Cannabis.* Nih.gov; National Academies Press (US). https://www.ncbi.nlm.nih.gov/books/NBK425751/

Anderson, L. (2014, May 18). *Marijuana: Effects, Medical Uses and Legalisation.* Drugs.com; Drugs.com. https://www.drugs.com/illicit/marijuana.html

Bridgeman, M. B., & Abazia, D. T. (2017). Medicinal Cannabis: History, Pharmacology, And Implications for the Acute Care Setting. *P & T: A Peer-Reviewed Journal for Formulary Management, 42*(3), 180–188. https://www.ncbi.nlm.nih.gov/pmc/articles/PMC5312634/

Farley, E., & Orchowsky, S. (2019, July). *Measuring the Criminal Justice System Impacts of Marijuana Legalisation and*

Decriminalisation Using State Data | Office of Justice Programs. Www.ojp.gov. https://www.ojp.gov/ncjrs/virtual-library/ abstracts/measuring-criminal-justice-system-impacts-marijuana-legalisation-0

Goldsmith, R. S., Targino, M. C., Fanciullo, G. J., Martin, D. W., Hartenbaum, N. P., White, J. M., & Franklin, P. (2015). Medical Marijuana in the Workplace. *Journal of Occupational and Environmental Medicine, 57*(5), 518–525. https://doi.org/10.1097/jom.0000000000000454

Grant, K. S., Petroff, R., Isoherranen, N., Stella, N., & Burbacher, T. M. (2018). Cannabis use during pregnancy: Pharmacokinetics and effects on child development. *Pharmacology & Therapeutics, 182*(5645664564646), 133–151. https://doi.org/10.1016/j.pharmthera.2017.08.014

History.com Editors. (2017, May 31). *Marijuana.* HISTORY; A&E Television Networks. https://www.history.com/topics/crime/history-of-marijuana

Krishna, M. (2022, November 16). *The Economic Benefits of Legalising Marijuana.* Investopedia. https://www.investopedia.com/articles/insights/110916/economic-benefits-legalising-weed.asp#:~:text=Increased%20tax%20revenues%2C%20job%20growth

Office of the Surgeon General, Assistant Secretary for Health (ASH. (2019, August 29). *Surgeon General's Advisory: Marijuana Use & the Developing Brain.* HHS.gov. https://www.hhs.gov/surgeongeneral/reports-and-publications/addiction-and-substance-misuse/advisory-on-marijuana-use-and-developing-brain/index.html

Popovici, I., & French, M. T. (2013). Cannabis Use, Employment, and Income: Fixed-Effects Analysis of Panel Data. *The Journal of Behavioural Health Services & Research, 41*(2), 185–202. https://doi.org/10.1007/s11414-013-9349-8

Renard, J., & Konefal, S. (2020). *Key Points Clearing the Smoke on Cannabis Use During Pregnancy and Breastfeeding -An Update.* https://ccsa.ca/sites/default/files/2022-05/CCSA-Cannabis-Use-Pregnancy-Breastfeeding-Report-2022-en.pdf

Roncero, C., Valriberas-Herrero, I., Mezzatesta-Gava, M., Villegas, J. L., Aguilar, L., & Grau-López, L. (2020). Cannabis use during pregnancy and its relationship with fetal developmental outcomes and psychiatric disorders. A systematic review. *Reproductive Health, 17*(1). https://doi.org/10.1186/s12978-020-0880-9

SAMHSA. (2019, March 25). *Know the Risks of Marijuana | SAMHSA - Substance Abuse and Mental Health Services Administration.* Samhsa.gov; SAMHSA. https://www.samhsa.gov/marijuana

Stohr, M., Willits, D., Makin, D., Hemmens, C., Lovrich, N., Stanton, D., & Meise, M. (2020). *Effects of Marijuana Legalisation on Law Enforcement and Crime: Final Report.* https://www.ojp.gov/pdffiles1/nij/grants/255060.pdf

Svrakic, D. M., Lustman, P. J., Mallya, A., Lynn, T. A., Finney, R., & Svrakic, N. M. (2012). Legalisation, decriminalisation & medicinal use of cannabis: a scientific and public health perspective. *Missouri Medicine, 109*(2), 90–98. https://www.ncbi.nlm.nih.gov/pmc/articles/PMC6181739/

THE UNIVERSITY OF SYDNEY. (2017). *History of cannabis.* The University of Sydney. https://www.sydney.edu.au/lambert/medicinal-cannabis/history-of-cannabis.html

UNODC. (2018). *4 DRUGS AND AGE Drugs and associated issues among young people and older people.* https://www.unodc.org/wdr2018/prelaunch/WDR18_Booklet_4_YOUTH.pdf

Wilkinson, S. T. (2013). Medical and recreational marijuana: commentary and review of the literature. *Missouri Medicine, 110*(6), 524–528. https://www.ncbi.nlm.nih.gov/pmc/articles/PMC6179811

Chapter 1: Similarities to Fetal Alcohol Spectrum Disorder

Noah Serethe

Introduction

Cannabis is the illegal substance that pregnant women use the most frequently. Prenatal cannabis exposure changes brain development and may have long-lasting effects on cognitive abilities, according to human epidemiological research and animal studies. We have learned a lot about the physiological functions of endogenous ligands (endocannabinoids) and their receptors as a result of research into the therapeutic potential of cannabis-based medications and synthetic cannabinoid molecules. Many different developmental issues in children have been linked to mother cannabis usage during pregnancy, according to studies. They can include underweight at birth, neurobehavioral development that is compromised, and a higher chance of developing attention-deficit/hyperactivity disorder (ADHD) and other behavioural issues. FASD is more common among certain populations, including "children in foster care, those with parents or caregivers who abuse alcohol or drugs, and those living in poverty" (American Academy of Pediatrics, 2015). For example, studies have indicated that cannabis usage during pregnancy can affect the baby's brain's development. The structure and operation of the brain's reward system may have changed as a result of these changes, which may increase the risk of addiction in later life.

Significant brain development problems are linked to both FCS and FASD, and these problems can have an impact on cognitive function, learning, memory, and social skills. Fetal Alcohol Syndrome (FASD) is "a pattern of mental and physical defects that can develop in a fetus when a woman drinks alcohol during pregnancy" (Mayo The clinic, 2021) Regarding how each illness affects the fetal brain's development, there are some significant variances between the two. The term "fetal cannabis syndrome" (FCS) is used to refer to a collection of developmental problems that may manifest in fetuses whose mothers consumed cannabis while they were pregnant. Fetal alcohol spectrum disorder (FASD), which is brought on by mother alcohol use during pregnancy, and this condition share many similarities. On the other hand, maternal alcohol usage during pregnancy is the cause of FASD. Alcohol can pass the placental barrier, just like cannabis, and have an impact on a fetus's brain development. The time and amount of alcohol consumption, genetics, and environmental circumstances, among other things, can all have a significant impact on how severe the illness is. Many developmental issues, such as facial defects, growth retardation, cognitive impairments, and behavioural issues, can be brought on by FASD. Alcohol is thought to have these effects in part because it can obstruct the fetal brain's proper neuronal growth.

Amount and Pattern of Alcohol Use on Neurobehavioral Outcome

The effects of maternal alcohol use on the neurobehavioral outcome of the fetus rely on many variables, including the quantity and pattern of alcohol use as well as the timing of exposure during development. It is estimated that FASD affects between 1% and 5% of children in the United States (American Academy of Pediatrics, 2015). The severity of the consequence is typically connected with the amount of alcohol ingested. However, how alcohol is consumed can frequently negate these effects, with binge-like consumption leading to worse deficits than chronic consumption. The timing of the exposure concerning development is crucial. The pattern and severity of anatomical and functional defects can be dramatically influenced by alcohol exposure at various stages of fetal development. This variance in the effects of maternal drinking may also be influenced by numerous other factors. For instance, consider the impact of a prenatal mother who consumes alcohol. This can affect elements comprising, but not limited to: maternal metabolism, genetic susceptibility, and susceptibility of various brain regions. However, the prevalence and severity of alcohol-induced developmental brain problems are most likely to be influenced by factors that change the peak blood alcohol concentration (BAC) that the embryo or fetus experiences (for example, maternal drinking habits and maternal metabolism). Unfortunately, it can be challenging to record this level of specificity, especially in populations that were recruited retrospectively, and different research provides different levels of information about exposure levels and patterns. These issues highlight the modifying variables that contribute to the considerable range of phenotypes presented in children with **FASD.**

Neurobehavioral Abnormalities in Children with FASD

As the biggest preventable cause of mental impairment, FASD affects an estimated 2%–5% of children in the United States and other Western European nations. Data from a US national poll showed that 13% of women continue to drink alcohol during pregnancy, despite public relations campaigns urging women to cut back once they learn they are pregnant. The symptoms of FASD can include "facial

abnormalities, small head size, shorter-than-average height, low body, poor coordination, and hyperactivity (Mayo Clinic, 2021). General intelligence, memory, language, attention, learning, visuospatial abilities, executive functioning, fine and gross motor skills and social and adaptive functioning are just a few of the areas where extensive cognitive abnormalities have been linked to

prenatal alcohol exposure. Although children with a diagnosis of FASD tend to have more severe impairments compared to those who were exposed to alcohol during pregnancy but do not have enough dysmorphic traits for a diagnosis, many FASD persons also show intellectual impairment and growth retardation. Intellectual disability is not a requirement for the diagnosis of FASD, and not all people with FAS are intellectually handicapped (IQ score 70). For children with FASD, the estimated average IQ is 70, whereas for those without dysmorphic disorders, it is 80. The outcomes of the few studies that have been done on those who have moderate alcohol intake in terms of intellectual ability have been mixed. The discrepancies among studies with moderate levels of exposure warrant the need for further research and the importance of considering a variety of factors as described in the proceeding section when evaluating studies.

Learning and memory deficiencies in kids who had substantial prenatal alcohol exposure have also been documented by clinical trials. Children exposed to prenatal alcohol have worse academic success and more learning disabilities than control children, which may be related to problems with verbal and nonverbal learning and memory. Both children with and without the dysmorphic characteristics of FAS have these deficiencies. According to certain research, children who have been exposed to alcohol retain linguistic information over the long term, but the initial encoding processes may be compromised. Children with FASD demonstrated visuospatial processing deficiencies in addition to difficulties with nonverbal memory, indicating damage and anomalies in the frontal-subcortical system. Children with FASD commonly had focus problems and attention deficit hyperactivity disorder (ADHD). Executive function and spatial processing may be particularly vulnerable to alcohol intake during pregnancy, according to a recent collaborative study (Johnson, 2018). The frontal-subcortical circuits, which involve projections from the frontal lobes to the basal ganglia and thalamic nuclei, control executive processes. These regions are altered by prenatal alcohol consumption. Children with FASD also exhibit social and adaptive deficiencies because they are more likely than non-exposed children to be labelled as hyperactive, disruptive, impulsive, or delinquent. Also, children exposed to alcohol during pregnancy have trouble resolving disputes and foreseeing the results of their behaviour. Also, children exposed to alcohol during pregnancy have trouble resolving disputes and foreseeing the results of their behaviour.

Neurodevelopmental Outcomes of Prenatal Cannabinoid Exposure

Over the past two decades, increasing epidemiological and experimental evidence has shown a link between cannabis use in adolescence (a crucial time for neurodevelopment) and an increased risk of cognitive impairments and neuropsychiatric illness. Moreover, the ECS plays a crucial role in synaptic plasticity, neuronal cell proliferation, and differentiation during embryonic neurodevelopment. Since that 9-THC easily passes the placental barrier from the mother to the fetus, in utero cannabis exposure may harm these processes and the long-term cognitive results.

The Ottawa Prenatal Prospective Study (OPPS), the Maternal Health Practices and Child Development Study (MHPCD), and the Generation R Study are the three largest prospective longitudinal cohorts that have been used to date to examine the effects of prenatal cannabis exposure on neurodevelopment (GenR). These findings show how cannabis consumption during pregnancy affects cognitive and behavioural areas. The Bayley Scale of Infant Development (BSID) scores, as well as increases in hyperactivity, impulsivity, and aggression, were all negatively impacted by cannabis exposure throughout childhood and adolescence. At age 10, exposure in the MHPCD cohort was also associated with worse reading and spelling scores on the Wide Range Achievement Test-Revised (WRAT—R), a test of academic achievement. The statistical differences were not clinically significant, according to this review's findings as well. However, it's significant to note that these data come primarily from the 3 big prospective studies, all of which began between 1978 and 2001. Consequently, the rapid increase in 9-THC concentrations reported over the previous 20 years and recent trends towards cannabis legalization, together with a higher frequency of use, are essentially unaccounted for in our analyses.

Furthermore, in a more recent retrospective observational cohort, a positive mother Δ9-THC urine test at the first prenatal visit was associated with aberrant 12-month developmental scores in children, as measured by the Ages and Stages: Social-Emotional Questionnaire (ASQ-SE). Furthermore, after controlling for potentially confounding covariates, recent cross-sectional findings from the ongoing Adolescent Brain Cognitive Development (ABCD) study, which enrolled 11,875 kids between the ages of 9 and 11 years, revealed that prenatal exposure to cannabis was connected to deficits in attention thought, and social issues (Balapal, 2015). Prenatal exposure to alcohol is the leading preventable cause of birth defects and intellectual disabilities in the United States (National Organization on Fetal Alcohol Syndrome, 2021) A significant retrospective examination of infants born between 2007 and 2012 Ontario, Canada, also found a moderate rise in the frequency of intellectual disability and learning problems, albeit these findings lacked statistical support (85). Mechanistic plausibility also supports the typically mild cognitive impairments seen in longitudinal cohorts.

Paternal Cannabinoid Use and Epigenetic Considerations

The maternal environment has been the primary focus of the vast majority of studies looking at the long-term impacts of exosomes in pregnancy to date. As previously mentioned, analysis of the GenR cohort showed that independent of maternal cannabis use, father cannabis use was predictive of psychotic-like experiences and behavioural impairments in offspring at ages 7 to 10. In this cohort, it is noteworthy because information about the father's cannabis usage was acquired from the mother's reports and was only recorded during pregnancy, not before. It is plausible that father preconception cannabis use is causally related to the observed offspring phenotypes, despite the authors of this study's hypothesis that there may be a similar aetiology underpinning both parental cannabis use and offspring behavioural outcomes. 9-THC exposure during adolescence, before mating, has recently been shown to affect the neurodevelopmental and behavioural consequences in later generations in rats. The heightened heroin self-administration in the offspring of parents who had been exposed to 9-THC but who had not themselves been exposed was accompanied by molecular and electrophysiological changes in the striatum, a crucial part of the reward circuitry. Moreover, sex-specific impacts were seen at both the gene expression and behaviour levels.

Regarding specific paternal cannabis use, there is no concrete evidence in both humans and rats showing that exposure to 9-THC changes the DNA methylation in sperm cells. These changes might serve as a vehicle for the effects of paternal toxicant exposure on the offspring's development and genetic expression. It has been demonstrated that adult male children of premating 9-THC exposed dads have deficiencies in attentional performance and memory tasks, along with changes in cholinergic signalling. It's interesting to note that THC exposure throughout pregnancy and adolescence has been demonstrated to significantly increase DA pathway sensitivity, with effects that last into adulthood. Such disruption of mesocortical and mesolimbic DAergic transmission patterns brought on by 9-THC may serve as important biomarkers for higher addiction risks as well as an underlying mechanism connected to greater susceptibility to schizophrenia, mood, and anxiety disorders. Interestingly, the data on the impact of father cannabis usage on the outcomes of offspring is still in its infancy and is primarily preclinical.

Role of the Endocannabinoid System during Brain Development and in Ethanol-Induced Brain Abnormalities

Due to the development and identification of the receptors that bind delta-9-tetrahydrocannabinol (9-THC) and their endogenous ligands, endocannabinoids (ECs), in animal tissues, cannabinoid research has gained a great deal of attention in the past twenty years (for a recent review see). The EC system regulates synaptic neurotransmission in many brain regions of the developing and adult brain in a variety of ways, according to the growing corpus of research. Increasing research has shown that EC signalling plays important role in the molecular networks that underlie both transient and persistent changes in synaptic strength. The crucial role that ECs play in some synaptic neurotransmission systems may alter the viability of the existing cellular theories of learning and memory.

Role of Cannabinoid Receptors and their Signalling

Since the 1980s, there has been evidence supporting the existence of a marijuana receptor. It has recently been established that cannabinoids comprise two distinct, cloned receptor subtypes known as CB1 and CB2 (Clarke, 2016) The literature has documented evidence for a third receptor (also known as the "AEA receptor") in brain and endothelial tissues. Nonetheless, CB3 has not yet been cloned, expressed, or characterized. The G protein-coupled receptors (GPCRs) CB1 and CB2 link to Gi/o proteins and are members of a broad superfamily of heptahelical G protein-coupled receptors (for more details, see reviews). Sometimes referred to as the "brain cannabinoid receptor," the CB1 receptor is mostly expressed in the brain and spinal cord. The expression levels of CB1 receptors are comparable to those of GABA- and glutamate-gated ion channels, making them among the most prevalent GPCRs in the brain. Throughout the past few years, there has been a lot of debate regarding the CNS's functional CB2 receptors. It is now acknowledged that CB2, which was once thought to be a receptor that was only found in the peripheral nervous system and was frequently called a "peripheral cannabinoid receptor," is also present in the brains of several animal species, including humans, in specific locations and small quantities (Balapal, 2015). On the other hand, the CNS receptor's functional significance is only now beginning to become clear.

Throughout the growing neurological system, the CB1 receptor has a diverse pattern of expression, and from the embryo's earliest stages on, its expression is correlated with the differentiation of neurons. The distribution of CB1 receptors in the fetal and neonatal rat brain, as well as the pattern of CB1 receptor mRNA expression has been characterized in numerous studies. Rats' GD 14—the moment at which the majority of neurotransmitters first show phenotypic expression—is when CB1 receptor mRNA levels and receptor binding may be seen. Because they are already connected to GTP-binding proteins at this stage of development, CB1 receptors seem to be functioning. CB1 receptors are present in greater quantities in the growing human and rat brains than in the adult brain. When compared to the adult brain, the distribution of CB1 receptors in the fetal and early neonatal brain is abnormal, especially in the forebrain's subventricular zones and white matter regions (Jones, 2017). Because the CB1 receptors are gradually acquired during the course of late postnatal development, the unusual position of these receptors is only a temporary event. This traditional pattern of distribution is then exhibited in the adult brain. In contrast to the adult brain, the fetal and early neonatal brain has an unusual distribution of CB1 receptors, especially in the forebrain's subventricular zones and white matter regions. Because the CB1 receptors are gradually acquired during the course of late postnatal development, the unusual position of these receptors is only a temporary event. This traditional pattern of distribution is then exhibited in the adult brain. The existence of CB1 receptors during early brain development suggest the potential involvement of CB1 receptors during fetal and early postnatal periods in specific events of CNS development, such as cell proliferation and migration, axonal elongation and, subsequently, synaptogenesis and myelinogenesis. Therefore, CB1 receptors play a role in the regulation of neurotransmitter release.

There is proof that cannabinoids during pregnancy affect how neurotransmitter systems mature and the behaviours they influence. CB1 receptors, which first appeared in the developing brain was responsible for these effects. A chronologically organized series of events that take place throughout the early stages of CNS development leads to a particular neurotransmitter's activity in the adult brain. Several of the functions connected to this neurotransmitter may change as a result of the dysregulation of this pattern. For instance, a change in the expression of the genes involved in the synthesis of receptors at a very precise developmental stage may affect certain actions connected to the physiological functions of these receptors. These changes could potentially result from a change in their concentrations or the activity of the signalling pathways associated with the CB1 receptors, as well as from an increase or reduction in their levels. It was shown that administering cannabinoids at dosages comparable to those observed in marijuana users altered the normal development of neurotransmitters, which most likely resulted in neurobehavioral abnormalities. As a result, adult animals exposed to cannabinoids during perinatal development display a variety of symptoms, including long-term changes in male copulatory behaviour, open-field activity, learning ability, stress response, pain sensitivity, social interaction and sexual motivation, drug-seeking behaviour, and neuroendocrine disturbances (Elizabeth, 2000). However, the majority of these neurobehavioral effects were induced by changes in the development of various neurotransmitter systems brought on by cannabis, most likely as a result of the activation of CB1 receptors during crucial prenatal and perinatal periods.

Conclusion

Fetal alcohol syndrome (FAS) and fetal cannabis syndrome (FCS) are both serious conditions that can have lifelong consequences for affected individuals. FAS occurs when a pregnant woman drinks alcohol, which can lead to physical, cognitive, and behavioural impairments in the developing fetus. FCS,

on the other hand, occurs when a pregnant woman uses cannabis, which can also lead to a range of developmental issues in the fetus. Both FAS and FCS can cause significant harm to the developing brain and body of a fetus, and the effects of these conditions can last a lifetime. However, Prevention of FASD

involves avoiding alcohol consumption during pregnancy and supporting women who may struggle with alcohol use through education and access to resources (CDC,2020). Although there is no cure for FASD, early intervention and supportive services can improve outcomes for affected individuals (Mayo Clinic, 2021). Education and awareness about the risks associated with alcohol and cannabis use during pregnancy are essential to prevent FAS and FCS. Healthcare providers should guide pregnant women and their partners about the risks of alcohol and cannabis use and the importance of prenatal care. In conclusion, FAS and FCS are preventable conditions that can cause significant harm to develop fetuses. Pregnant women and their partners should be informed about the risks of alcohol and cannabis use during pregnancy and encouraged to seek prenatal care. By working together to raise awareness and prevent these conditions, we can ensure healthier outcomes for mothers and their children.

References

Basavarajappa, B. S. (n.d.). *Fetal Alcohol Spectrum Disorder: Potential Role of Endocannabinoids Signaling*. MDPI. Retrieved March 7, 2023, from https://www.mdpi.com/2076-3425/5/4/456#B251-brainsci-05-00456

Cleveland Clinic. (2022, February 16). *Fetal Alcohol Syndrome (FAS): Symptoms, Causes & Treatment*. Retrieved March 8, 2023, from https://my.clevelandclinic.org/health/diseases/15677-fetal-alcohol-syndrome

De Rose, C. (2019, December 19). *Two case reports of fetal alcohol syndrome: broadening into the spectrum of cardiac disease to personalize and to improve clinical assessment - Italian Journal of Pediatrics*. Italian Journal of Pediatrics. Retrieved March 8, 2023, from https://ijponline.biomedcentral.com/articles/10.1186/s13052-019-0759-y

Gavin, M. L. (n.d.). *Fetal Alcohol Syndrome (for Parents) - Nemours KidsHealth*. Kids Health. Retrieved March 8, 2023, from https://kidshealth.org/en/parents/fas.html

NCBI. (2021, February 12). *Use of Cannabis in Fetal Alcohol Spectrum Disorder*. Retrieved March 7, 2023, from https://www.ncbi.nlm.nih.gov/pmc/articles/PMC7891191/

Psychiatric Times. (2020, October 9). *Cannabis in Pregnancy – Rejoinder, Exposition and Cautionary Tales*. Retrieved March 8, 2023, from https://www.psychiatrictimes.com/view/cannabis-pregnancy-rejoinder-exposition-cautionary-tales

Rubia, K. (2020, December 14). *Prenatal Cannabinoid Exposure: Emerging Evidence of Physiological and Neuropsychiatric Abnormalities*. Frontiers. Retrieved March 7, 2023, from https://www.frontiersin.org/articles/10.3389/fpsyt.2020.624275/full

Simmons, B., & Sharpe, H. (2021, December 31). *Responding to the Unique Complexities of Fetal Alcohol Spectrum Disorder*. Frontiers. Retrieved March 8, 2023, from https://www.frontiersin.org/articles/10.3389/fpsyg.2021.778471/ful

Chapter 2: Interactions with the Fetal Brain: Examining the Specific Ways in which Cannabinoids Interact with Fetal Brain Development and the Potential for Damage

Eline El-Awad Gonzalez

Introduction

The increase in the current therapeutic potential of cannabinoids has provided a large novel insight into the physiological implications of endogenous ligands and their respective receptors. Herein, this chapter will explore the influence of prenatal exposure to cannabinoids on fetal brain development and the potential for damage, followed by an overview of the molecular composition of cannabinoid systems involved in biomolecular interactions. Ultimately, a fundamental understanding of how cannabinoid signalling modulates foundational developmental processes such as cell proliferation, neurogenesis, migration and axonal pathfinding is important in understanding the safety of cannabis use during pregnancy.

The use of cannabis during pregnancy can have negative effects on prenatal development (Friedrich et al., 2016). Cannabis exposure during pregnancy can result in reduced birth weight, increased risk of developmental delays and behavioural problems, and increased risk of certain birth defects. Additionally, such use can also increase the risk of stillbirth and preterm labour. It is therefore often recommended that pregnant women avoid using cannabis to protect the health and development of their fetus.

THC, the principal psychoactive component in cannabis, is a lipophilic compound capable of crossing the blood-brain barrier and the placenta (Martin et al., 1997). A primate study has revealed that THC is detectable in fetal blood in under 15 minutes following an intravenous infusion in the mother. Similar studies conducted in canines found that 30% of maternal plasma levels in fetal fat were composed of THC following injection. Importantly, it was determined that when cannabis use is *occasional* during pregnancy, the fetal exposure to TLC continued for the entirety of the development timeframe in the womb (Psychoyos et al., 2008).

Cannabinoids primarily interact with two prominent G protein-coupled receptors, CB1 and CB2. While CB2 receptors are sparse and found in immune and neuronal cells, CB1 receptors are present in high quantities in gonads and the brain (Mackie, 2008). When taken, exogenous cannabinoids disrupt the tightly regulated signalling pathway mediated by endogenous cannabinoids produced by the human body. Such disruption may have implications in downstream effects on obstetrical outcomes and embryo development.

Currently, there are five well-characterised ways in which cannabinoids affect obstetrical outcomes. Implantation, the attachment of the fertilised egg to the uterine wall, is dependent upon several tightly regulated processes. Here are a few ways this can happen:

1. **Impairment of Fallopian Motility:** Cannabinoid signalling controls muscle contraction and relaxation in the fallopian tube responsible for the movement of a fertilised egg through the fallopian tube and into the uterus.

2. **Ectopic Pregnancy:** Previous studies have shown that the blood collected from women who have had

ectopic pregnancies contains significantly higher levels of the naturally occurring cannabinoid, anandamide, compared to normal pregnant controls. Consistency between human and animal data adds confidence that the observed findings in animal models of altered cannabinoid signalling may play a role in ectopic pregnancy.

3. **Non-Hatched or Non-Viable Embryo:** In mice models known to have altered cannabinoid signalling, increased mortality of offspring was observed in association with implantation of slowly developing embryos.

4. **Decreased Uterine Receptivity:** It is theorised that the binding of exogenous cannabinoids to CB1 receptors in the uterus has embryotoxic effects on the uterine environment. Modelling of this scenario has halted the development of blastocysts in vivo and in vitro.

5. **Miscarriage (Spontaneous Abortion):** Folic acid (Vitamin B9) is essential for embryo development and cannot be synthesised by the body which is why women are encouraged to take folic acid supplements during pregnancy. THC significantly decreases fetal folic acid uptake. Low levels of folic acid during pregnancy are associated with higher rates of miscarriages, as well as neural tube defects and low birth weight.

Tetrahydrocannabinol (THC) – Active Cannabis Agent

When affecting embryo development, THC crosses the placenta, enters fetal blood circulation, passes through the blood-brain barrier, and is found at the highest levels in fetal fat tissue. The brain is 60% fat and therefore stores THC following maternal ingestion (Paria et al., 2001). The brain is also densely populated with CB1 receptors which mediate THC's psychoactive properties.

1. **Folic Acid Uptake:** As stated above, THC interferes with fetal folic acid uptake. Low levels of folic acid during pregnancy are known to be associated with neural tube defects and low birth weight.

2. **Cellular Growth:** Exogenous cannabinoids may interfere with critical pathways for cellular growth and angiogenesis (formation of new blood vessels).

3. **Neural Development:** Cannabinoids acting upon the CB1 receptor can influence the differentiation of neural cells from stem cells in the brain. This has tremendous potential to negatively affect learning and memory as well as developmental processes such as limb development.

Although the pharmacokinetic profiles of cannabinoids' active compounds (such as THC) vary depending on users, dosage, route of administration, etc., the bioavailability ranges from 25% to 31% (Casu et al., 2005). Bioavailability can be described as the ability of a drug to be absorbed and used by the body; in this context, such percentages are quite high of cannabinoids. Distribution of THC is time-dependent and begins immediately upon absorption into fatty tissues and highly perfused organs such as the brain, heart, lung, and liver. Initially, THC has a rapid half-life of ~6 minutes, however, the long terminal half-life can last up to 22 hours.

Concerning metabolism, exogenous cannabinoids follow a different trajectory than endogenous cannabinoids (Hurd et al, 2001). The metabolic route, in this case, is hypothesised to involve cytochrome P450 oxidases (CYP). THC is metabolised to about 80 human metabolites, mainly by CYP2C9, -2C19, and -3A4. Some of the metabolites are active. The metabolism of other cannabinoids has been less studied but likely proceeds through similar reactions. In addition to the above enzymes, CYP2D6 is also known to be inhibited by CBD, CBN, and THC. Among these, CBD is generally the most potent inhibitor of the various CYPs, which include CYP2C8, -2C9, and -3A4, the main enzymatic pathways relevant to the most commonly used health products. Carboxylesterases CES1 and CES2 were found to metabolise two of these substances or their derivative. It has been noted that pre-administration of CBD may potentiate the effect of THC, possibly by decreased metabolism of the THC. Specifically, smoked cannabis likely induces CYP1A2-mediated theophylline metabolism. Some THC isomers and CBD exert no or minimal impact on platelet monoamine oxidase activity ex vivo, although some cannabis substances may inhibit the monoamine oxidase-B isoform. Chronic cannabis smoking did not affect ex vivo plasma dopamine b-hydroxylase activity acutely.

Cannabinoids cross the placental barrier (organ linking the mother and fetus) and are known to be present in breast milk. CBD can inhibit the main fetal drug disposition enzyme CYP3A7. CBN and THC affect this enzyme to a lesser extent. However, the inhibitory potential of these substances on other fetal enzymes such as flavin-containing monooxygenases is not known. Interestingly, a study was conducted on 42 brain samples from singleton abortions (Wang et al., 2003). The fetuses were at the midgestational stage of development (~20 weeks). The mothers were predominantly young African-American women with an average age of 23 years. The investigation on the effects of in-utero cannabis exposure on mRNA expression was focused on the caudal level of the fetal forebrain in which the most intense CB1 expression was located. The data provided evidence for impairments of the endogenous opioid system in association with in-utero cannabis exposure in the human fetal brain. Maternal cannabis use was significantly associated with reduced mRNA expression levels for PENK in the putamen (both patch and matrix compartments), increased mu expression in the amygdala and decreased kappa expression in the mediodorsal thalamic nucleus. After controlling for various confounding factors, the data revealed specific associations between cannabis exposure during pregnancy and opioid gene expression (Richardson et al., 1989).

A number of the present human results are consistent with observations obtained with animal models. Rat studies have documented that fetal animals exposed to THC in utero have decreased PENK mRNA expression levels in the caudate-putamen. PR proenkephalin mRNA levels in the caudate-putamen are also significantly reduced when perinatally exposed rats are studied at an adult age. However, the effects of prenatal THC treatment on PENK expression are strongly dependent on sex. Significant alterations were observed only in fetal male and adult female rats. Alterations of PENK expression levels are also evident in rats that receive chronic THC administration during adulthood (Dreher et al., 1994). Taken together, it is apparent that disturbances of PENK gene expression are a common consequence of repeated exposure to THC in rat models though the exact mRNA changes are dependent on the pattern of drug administration, dose and time course of drug administration, brain regions examined, age when the rat is studied and the sex.

Fetal Brain Implications

The specific alterations observed for the opioid gene expression in the fetal brain in association with drug exposure are of significant relevance since the opioid system plays a critical role in the regulation of emotions, reinforcement, cognition, motor function and nociception (Richardson et al., 2002). The mediodorsal thalamus is a key component of the limbic system connecting subcortical structures such as the nucleus accumbens and amygdala with the prefrontal cortex. The amygdala is a critical limbic brain structure for the regulation and expression of emotions. The amygdala also affects higher cognitive functions through its reciprocal connections with the prefrontal cortex (decision-making, working memory) and hippocampus (memory consolidation). Moreover, the amygdala has strong connections with the ventral tegmental area, the origin of the mesocorticolimbic system, and the nucleus accumbens which are central to the reward behaviour (Fried et al., 1988). Thus, disturbances of amygdala dopamine- and opioid-related function during development could influence cortical and limbic system functions. Longitudinal human studies have documented that prenatal marijuana-exposed children and young adults exhibit a variety of emotional, cognitive and behavioural deficits, such as deficit executive function, depression, anxiety, inattention and delinquency (Fried et al., 1990).

A central role for cannabinoid signalling in brain development is now emerging. Perinatal and adolescent cannabis exposure may disrupt the precise temporal and spatial control of cannabinoid signalling at critical stages of neural development, leading to detrimental effects on later nervous system functioning (Day et al., 1994). Indeed, longitudinal studies in humans with prenatal cannabis exposure demonstrated exaggerated startle response and poor habituation to novel stimuli in infants, and hyperactivity, inattention and impaired executive function in adolescents (Fried et al., 2002). Many of these behavioural effects have also been modelled in animal studies. Furthermore, possible teratogenic effects of endocannabinoid system-based therapies in pregnant women and long-term exposure to cannabinoid signalling-modifying agents such as organophosphate pesticides need to be taken into consideration.

The Ottawa Prenatal Prospective Study (OPPS) found that prenatal marijuana exposure was highly correlated with an increase in exaggerated startles and tremors as well as with a significant reduction in habituation to light at the neonatal stage (Zammit et al., 2009). Altered sleep patterns were found in the Maternal Health Practices and Child Development Study (MHPCD), and the authors also reported a non-significant trend towards increased irritability. A study on neonates from adolescent mothers found in cannabis-exposed infants transiently increased irritability, excitability and arousal 24–72 h after birth (Stone et al., 2010). However, these symptoms were not reported within the MHPCD cohort or in an ethnographic field study based in Jamaica. The MHPCD cohort also demonstrated that a higher amount of cannabis use per day (defined as more than one joint per day) during the third trimester of pregnancy was associated with decreased mental scores of the Bayley Scales of Infant Development at 9 months of age, a difference that disappeared by 18 months. No cognitive deficit was observed during early childhood in the OPPS study, particularly between the ages of 1 and 3 years, suggesting that CNS abnormalities might be absent or subclinical in toddlers.

Upon closer inspection, the impact of prenatal marijuana exposure is a little more difficult to discern (Stone et al., 2010). For example, one report from the MHPCD cohort found that heavy first- and third-trimester exposure (rated as >0.89 joints/day) was associated with increased hyperactivity and impulsivity, while another found that heavy second-trimester exposure was significantly associated with increased impulsivity. First- and third-trimester exposure also predicted increased levels of depressive symptoms, assessed by the Children's Depression Inventory, whereas second-trimester usage was associated with some depressive, but fewer internalising, symptoms compared with the extent observed in first- and third-trimester exposure groups (Zammit et al., 2009). Verbal IQ, reading comprehension, overall IQ, presence of psychotic symptoms and sleep patterns do not seem to be impacted. A recent study has assessed volumetric changes using functional MRI (fMRI) in the brains of children exposed to some drugs, including marijuana, during pregnancy. This study found evidence of reduced cortical grey matter and parenchymal volume in children (aged 10- to 14 years old) with intra-uterine marijuana exposure.

Statistics on Cannabis Use and Implications

Cannabis use during 2010 for those aged 15–64 was estimated to be between 2.9 and 4.3% worldwide, with a high but steady occurrence in North America and Western/Central Europe (Psychoyos et al., 2008). The Substance Abuse and Mental Health Services Administration estimates that 7.1% of pregnant women aged 18–25 have used illicit drugs in the month before being surveyed. Marijuana was the most prevalent substance abused, ranging from 2–6% usage as determined by interviews or self-report. However, one study on cannabis usage during pregnancy found an 11% usage rate by measuring serum metabolites close to that seen in age-matched, non-pregnant women (10.9%). Moore et.al. found that within a British population, marijuana was the only illicit drug pregnant women were likely to continue using to term.

Available data linking prenatal cannabis exposure to congenital anomalies or preterm delivery is weak. While fetal alcohol syndrome-like features in prenatally cannabis-exposed newborns have been reported, several other studies have failed to replicate such findings (Friedrich et al., 2016). Nevertheless, prenatal cannabis exposure is associated with fetal growth restriction and learning disabilities and memory impairment in the exposed offspring. The mean potency of cannabis preparations, in terms of contents of its psychoactive constituent, THC, has increased from 3.4% in 1993 to 8.8% in 2008 and can reach as high as 30% in certain hashish preparations. This fact is important since THC effects are dose-related and classical studies carried out in the 1970s used doses that reflected cannabis intake at that time. Key findings from human and animal studies regarding behavioural consequences of cannabis exposure during pregnancy and/or lactation will be summarised in the following section.

For 3–4-year-old children, prenatal marijuana exposure negatively affected the verbal and memory domains in both the OPPS and MHPCD studied groups (Casu et al., 2005). Cognitive development assessed by the Stanford-Binet Intelligence Scale demonstrated a negative association of short-term memory and verbal reasoning with first and/or second-trimester marijuana usage. Similarly, memory and verbal domains, measured by the McCarthy Scales of Children's Abilities, decreased with daily marijuana usage. However, composite intelligence scores in both studies were not impacted at this age by maternal marijuana use.

When children reach school age at around 5–6 years old, reports on the consequences of prenatal marijuana exposure begin to diverge (Wang et al., 2003). Exposed children from the OPPS cohort appear to have no memory deficits, while those from the MHPCD cohort report short-term memory deficits that correlate strongly with heavy second-trimester exposure. Cannabis-exposed children in the OPPS cohort scored significantly lower in tests for sustained attention, while those from the MHPCD group displayed increased attention (measured by fewer errors of omission in a continuous performance task) from second-trimester exposure. Both groups reported an increase in impulsive and hyperactive behaviours. Follow-up studies found that problems of depression, hyperactivity, inattention and impulsivity persist into the 9–12-year age range, raising speculation of deficits in higher cognitive processes such as executive function.

Conclusion

In summary, cannabis consumption during pregnancy has profound but variable effects on offspring in several areas of cognitive development (Richardson et al., 2002). Most of the information on the long-term consequences of prenatal exposure to cannabis comes from longitudinal studies of the OPPS and MHPCD cohorts. By comparing data from the cohorts, a pattern emerges where maternal cannabis use is associated with impaired high-order cognitive function in the offspring, including attention deficits and impaired visuoperceptual integration (Dreher et al., 1994). Genetic and environmental interactions may affect the extent of long-term neurobehavioural deficits resulting from prenatal exposure.

Recent advances in methodology in prenatal substance use research employ novel approaches to disentangle the exposure to substance effects from correlated risk factors. For example, in the prospective Generation R Study, where 7452 mothers were enrolled during pregnancy and information on substance use and ultrasound measures of fetal growth in early, mid- and late pregnancy were collected, information on paternal cannabis use was also included. Thus, maternal cannabis use during pregnancy was associated with growth restriction in mid and late pregnancy, and also with lower birthweight, while no such association was found for paternal cannabis use in the same period, demonstrating a direct biological effect of maternal intrauterine exposure to cannabis on fetal growth. Refined study designs and novel approaches will assist in confirming and extending the findings of associations between prenatal cannabis exposure and offspring outcomes.

Summary

Marijuana is the most prevalent illicit substance abused by pregnant women, with an incidence of 2–6% (determined by interview or self-report) and as high as 11 % by serum. The mean potency of marijuana preparations, in terms of Δ9-tetrahydrocannabinol content, has increased from 3.4% in 1993 to 8.8% in 2008, reaching 30% in some hashish preparations. Human marijuana consumption during pregnancy appears to have lasting effects on the child's higher cognitive function. A better understanding of the molecular framework of endocannabinoid signalling may contribute to defining the molecular changes underlying the neurobehavioural changes observed in the offspring of cannabis users and the neurodevelopmental impact of the therapeutic manipulation of the endocannabinoid system during pregnancy.

References

Casu M. A., Pisu C., Sanna A., Tambaro S., Spada G. P., Mongeau R. (2005). Effect of delta9-tetrahydrocannabinol on phosphorylated CREB in rat cerebellum: an immunohistochemical study. *Brain Research*, 1048, 41–47

Day N. L., Richardson G. A., Goldschmidt L., et al. (1994) Effect of prenatal marijuana exposure on the cognitive development of offspring at age three. *Neurotoxicology and Teratology*, 16(2),

169–175.

Dreher M. C., Nugent K., Hudgins R. (1994). Prenatal marijuana exposure and neonatal outcomes in Jamaica: an ethnographic study. *Pediatrics*, 93(2), 254–260.

Fried P. A., Watkinson B. (1988) 12- and 24-month neurobehavioural follow-up of children prenatally exposed to marihuana, cigarettes and alcohol. *Neurotoxicology and Teratology*, 10(4), 305–313.

Fried P. A., Watkinson B. (1990) 36- and 48-month neurobehavioural follow-up of children prenatally exposed to marijuana, cigarettes, and alcohol. *Infant Behavioural and Development*, 11(2), 49–58

Fried P. A. (2002) Conceptual issues in behavioural teratology and their application in determining long-term sequelae of prenatal marihuana exposure. *Journal of Child Psychology and Psychiatry*, 43(1), 81–102.

Friedrich J., Khatib D., Parsa K., Santopietro A., Gallicano G. I. (2016). The grass isn't always greener: The effects of cannabis on embryological development. *BMC Pharmacology and Toxicology*, 17(1), 45

Hurd Y. L., Suzuki M., Sedvall G. C. (2001). D1 and D2 dopamine receptor mRNA expression in whole hemisphere sections of the human brain. *Journal of Chemistry and Neuroanatomy*, 22, 127–137.

Mackie K. (2008). Cannabinoid receptors: where they are and what they do. *Journal of Neuroendocrinology*, 1, 10-4

Martin B. R., Dewey W. L., Harris L. S., Beckner J. S. (1997). 3H-delta9-tetrahydrocannabinol distribution in pregnant dogs and their fetuses. *Research Communications in Chemical Pathology and Pharmacology*, 17(3), 357-70

Paria B. C., Song H., Wang X., Schmid P. C., Krebsbach R. J., Schmid H. H., Bonner T. I., Zimmer A., Dey S. K. (2001). Dysregulated cannabinoid signaling disrupts uterine receptivity for embryo implantation. *Journal of Biology and Chemistry*, 276(23), 20523-8

Psychoyos D., Hungund B., Cooper T., Finnell R. H. (2008). A Cannabinoid analogue of 3H-delta9-tetrahydrocannabinol disrupts neural development in chick. *Birth Defects Research*, 83, 477-488

Richardson G. A., Day N., Taylor P. (1989) The effect of prenatal alcohol, marijuana, and tobacco exposure on neonatal behaviour. *Infant Behavioural and Development*, 12, 199–209.

Richardson G. A., Ryan C., Willford J., Day N. L., Goldschmidt L. (2002). Prenatal alcohol and marijuana exposure. Effects on neuropsychological outcomes at 10 years. *Neurotoxicology Teratology*, 24, 309–320.

Stone K. C., Lagasse L. L., Lester B. M., et al. (2010) Sleep problems in children with prenatal substance exposure: the Maternal Lifestyle study. *Archives of Pediatrics & Adolescent Medicine*, 164(5), 452–456.

Wang X., Dow-Edwards D., Keller E., Hurd Y. L. (2003) Preferential limbic expression of the cannabinoid receptor mRNA in the human fetal brain. *Neuroscience*, 118, 681–694.

Zammit S., Thomas K., Thompson A., et al. (2009) Maternal tobacco, cannabis and alcohol use during pregnancy and risk of adolescent psychotic symptoms in offspring. *British Journal of Psychiatry*, 195(4), 294–300.

Chapter 3: History of Cannabis Legalisation in Canada and Uruguay, and Exploring the Impact of Cannabis Legalisation Fetal Cannabis Syndrome Risk

Anoushka Kaliyambath

Introduction

Cannabis legalisation has been spreading globally post the 2010s with Uruguay being the first country to legalise non-medical, or recreational use in 2013 (Hammond et. al, 2020). After that major policy change, many countries have been reevaluating the dangers cannabis poses and have enacted a notable shift toward cannabis commercialisation. Legalising cannabis has been proven to have various socioeconomic benefits including reduced substance possession costs, opportunities for new government and private revenue, employment growth, and improved public safety through legal and responsible distribution (Kavousi et al., 2021). However, similar to any addictive substance it can still pose a risk to public health, and vulnerable groups.

One group, especially at risk of the effects of cannabis use, is pregnant women susceptible to fetal cannabis syndrome. While the adverse effects of Fetal Cannabis Syndrome still require more in-depth research current studies show that cannabis use during pregnancy can result in various issues with the baby. Some of the major fetal issues that can occur include low birth weight, premature birth, long-term cardiovascular issues, issues with brain development and a variety of mental health issues including learning and development disabilities, low IQ, hyperactivity, and difficulty with low attention spans (Society of Obstetricians and Gynaecologists of Canada, 2023).

This chapter aims to explore the history of cannabis legalisation in both regions of Uruguay and Canada. The policy history surrounding both states will additionally be used to consider if legalisation has had notable effects on the prevalence of Fetal Cannabis Syndrome.

Brief Overview of Cannabis

Before unpacking the history of cannabis legalisation in different regions it is important to outline what cannabis is.

Cannabis is a flowering plant that comes in three primary species: Cannabis sativa, Cannabis indica, and Cannabis ruderalis (Government of Canada, 2023). The different species have different physical characteristics and chemical profiles, resulting in varying effects when consumed. Sativa strains are known for their uplifting and energizing effects, while Indica strains are known for their relaxing and sedative effects. Ruderalis strains are less common and often used for breeding purposes. Cannabis can be consumed in various forms, including smoking, vaping, edibles, tinctures, and topicals (Government of Canada, 2023). As outlined these strains of cannabis can be used both recreationally, and medically and have been used long before modern-day legalisation.

The main active ingredient in cannabis, THC, interacts with the body's endocannabinoid system, which regulates various functions such as mood, appetite, and pain. THC binds to the cannabinoid receptors in the brain and nervous system, producing the characteristic effects of cannabis, such as euphoria, altered perception, and relaxation (Government of Canada, 2023). However, cannabis is not exclusively used for the psychoactive effects of THC. Cannabis also contains other cannabinoids and compounds that have been researched and historically considered to hold therapeutic benefits. One such cannabinoid is cannabidiol (CBD), which has been studied and utilised for its anti-inflammatory and anxiety-reducing properties (Government of Canada, 2023).

As outlined cannabis is not inherently dangerous however, particularly if used inappropriately, it can cause undesirable short-term and long-term effects, some of which may cause permanent alterations to the mind or body.

Uruguay History

During the colonial period, cannabis was widely cultivated and used in South America, particularly in the Andean region, where it was known as marijuana (Queirollo et al. 2018). It was used by indigenous peoples for its psychoactive properties and was also used in traditional medicine to treat a range of ailments.

In the late 19th and early 20th centuries, the use of cannabis in South America became more widespread and was embraced by artists, intellectuals, and bohemians. In countries like Argentina, Brazil, and Chile, cannabis-smoking clubs known as *fumaderos* became popular social gatherings (Queirollo et al. 2018). However, by the mid-20th century, attitudes towards cannabis began to shift. In the 1940s and 1950s, South American countries, under pressure from the United States, began to criminalise the possession, cultivation, and sale of cannabis. This was largely due to the influence of the United States anti-drug policies, which were driven by a moral panic surrounding drug use (Queirollo et al. 2018).

Despite the criminalisation of cannabis, its use continued to be widespread in South America, particularly among marginalised communities. In countries like Colombia and Peru, cannabis cultivation became a lucrative business for drug cartels, leading to a rise in violence and instability. However, attitudes towards cannabis have begun to shift once again. In many South American countries, including Uruguay, Argentina, and Colombia, cannabis has been decriminalised or legalised for medical or recreational use. This has been driven in part by a growing recognition of the potential medical benefits of cannabis, as well as a desire to reduce the harm caused by the war on drugs.

Uruguay, a small country located in South America, has been at the forefront of cannabis legalisation. In 2013, Uruguay became the first country in the world to legalise the production, sale, and consumption of cannabis for recreational purposes. This decision was not made overnight, and it was the result of a long process that began in the early 2000s. In this essay, we will explore the history of cannabis legalisation in Uruguay, the reasons behind the decision, and the impact of this policy change on the country and the world.

Uruguay's journey towards cannabis legalisation began in 1974, when the country's military dictatorship banned all drugs, including cannabis. This ban remained in place until the country returned to democracy in 1985. However, the ban did not prevent the spread of cannabis use in Uruguay, and by the 2000s, cannabis had become the most widely used illegal drug in the country (Queirollo et al. 2018). The government's approach to drug policy during this time was primarily focused on law enforcement and punishing drug users and sellers.

In the early 2000s, Uruguay began to shift its drug policy towards a more progressive and public health-oriented approach. This shift was driven by a growing awareness of the failure of the war on drugs and the negative consequences of drug prohibition. In 2000, Uruguay decriminalised the possession of small amounts of cannabis and other drugs, meaning that individuals caught with small quantities of drugs were not punished criminally but were instead referred to treatment programs (Queirollo et al. 2018). This was a significant departure from the country's previous approach to drug policy and represented a recognition of the need to address drug use as a public health issue rather than a criminal one.

The next step in Uruguay's journey towards cannabis legalisation came in 2011 when President José Mujica proposed a bill to legalise and regulate the production, sale, and consumption of cannabis (Veldman, 2021). The bill was based on the idea that cannabis use should be treated as a public health issue rather than a criminal one, and that the government should take control of the cannabis market to reduce the harms associated with drug prohibition. The bill was also driven by a desire to reduce drug-related violence and corruption, as well as to provide a source of revenue for the government (Queirollo et al. 2018).

The proposal faced significant opposition from both within and outside the country. Many critics argued that legalising cannabis would lead to an increase in drug use and addiction and that it would send the wrong message to young people. Others argued that the proposal was in violation of international drug treaties and that it would harm Uruguay's relations with other countries. Despite these objections, the bill was passed by the Uruguayan parliament in December 2013, making Uruguay the first country in the world to legalise cannabis for recreational use. The law allowed individuals to grow up to six plants at home, form cannabis clubs with up to 45 members, or buy up to 40 grams per month from licensed pharmacies (Veldman, 2021). The law also established a regulatory authority to oversee the production, sale, and distribution of cannabis and required users to register with the government to purchase cannabis.

The implementation of the law was not without its challenges. The government faced difficulties in setting up the regulatory framework and ensuring that the supply of cannabis was sufficient to meet demand. There were also concerns about the potential for increased drug use among young people, and some critics argued that the law was not being enforced effectively. Despite these challenges, there is evidence to suggest that the legalisation of cannabis has had a positive impact on Uruguay. The government has been able to regulate the cannabis market and ensure that users have access to high-quality, safe products. The law has also generated revenue for the government, which has been used to fund drug treatment programs and other public health initiatives.

Canada's History

Cannabis has been used by Indigenous people for centuries for medicinal, spiritual, and ceremonial purposes. However, the colonial history of Canada and the introduction of prohibitionist drug policies have had a significant impact on Indigenous cannabis use and access.

Indigenous cannabis use was widespread across Canada long before the arrival of Europeans. The plant was used for medicinal purposes to treat a range of ailments, from pain and inflammation to anxiety and depression. Additionally, cannabis was used in spiritual and ceremonial contexts. The plant was considered to have spiritual properties, and its use was seen as a means of connecting with the natural world. European colonialists in Canada also used hemp, a variety of the cannabis plant, for its strong fibres, which were ideal for making rope, clothing, and paper. The plant quickly became an important crop for early settlers, with hemp farming expanding throughout the 17th and 18th centuries (Tattrie and Yarhi, 2019).

Cannabis use for recreational purposes became more widespread in the 20th century, particularly after the 1960s counterculture movement. However, the legal status of cannabis has been a source of controversy in Canada, with the government taking various approaches over the years. In 1923, cannabis was added to the Schedule of the Narcotics Drug Act, making it illegal to possess or sell the drug in Canada (Tattrie and Yarhi, 2019). However, it wasn't until the 1960s and 1970s that cannabis use became more prevalent, particularly among young people. This led to calls for reform and a growing movement for the legalisation of cannabis.

In 2001, Canada became one of the first countries in the world to legalise the use of medical cannabis, with the passing of the Marijuana for Medical Purposes Regulations (MMPR) (Tattrie and Yarhi, 2019). The MMPR allowed patients with certain medical conditions to access cannabis for medical purposes with a doctor's prescription. This marked a significant shift in attitudes towards cannabis, with the recognition of its potential therapeutic benefits.

In 2018, Canada became the second country in the world to fully legalise the recreational use of cannabis, following Uruguay. The *Cannabis Act*, which came into effect on October 17, 2018, allowed adults over the age of 18 to possess and consume cannabis for recreational purposes. It also created a legal framework for the production, distribution, and sale of cannabis products, with strict regulations in place to ensure safety and quality (Tattrie and Yarhi, 2019).

Since the legalisation of cannabis, the industry has grown significantly, with a wide range of products now available to consumers. These include dried flowers, oils, edibles, and concentrates, among others. However, the legalisation of cannabis has also presented new challenges, particularly in terms of regulating the industry and ensuring public safety.

While there are still challenges facing the industry, particularly in terms of regulation and supply, the legalisation of cannabis has opened up new opportunities for businesses and consumers alike. With ongoing investment and development, the cannabis industry in Canada is poised for continued growth in the years ahead.

Dangers of Legalising Cannabis for Pregnant women

As touched on previously, pregnant women are especially susceptible to the negative, and dangerous effects of THC within the cannabis plant strains. The use of cannabis during pregnancy has been associated with adverse effects on fetal development. Both The Society of Obstetricians and Gynaecologists of Canada (2023), and The American College of Obstetricians and Gynecologists (2023) recommends that pregnant women should not use cannabis due to the potential risks to the fetus. However, despite these warnings, the use of cannabis during pregnancy has been increasing in recent years, particularly in countries where the drug has been legalised for recreational use.

Fetal Cannabis Syndrome was first formally researched and acknowledged by scientists in the late 1970s to early 1980s when researchers observed an increasingly prevalent set of physical and behavioural abnormalities in infants born to mothers who used cannabis during pregnancy (Wu et al., 2011). These infants had lower birth weights, smaller head circumferences, and reduced gestational age compared to infants born to non-using mothers. They also showed developmental delays in areas such as language, attention, and memory (Wu et al., 2011).

The exact mechanism by which cannabis causes these developmental problems is not well understood. However, it is believed that THC affects the developing brain by disrupting the endocannabinoid system, which is responsible for regulating various physiological processes such as mood, appetite, and sleep (Government of Canada, 2023). This disruption can result in abnormal brain development, leading to impaired cognitive and behavioural functioning.

One of the most common physical symptoms of Fetal Cannabis Syndrome is low birth weight. Babies born to mothers who use cannabis during pregnancy are often smaller and lighter than babies born to non-using mothers. Low birth weight is associated with a higher risk of health problems such as respiratory distress, infections, and developmental delays (Roncero et al., 2020). Babies born to cannabis-using mothers also have a higher risk of stillbirth, preterm birth, and neonatal intensive care unit admission (Society of Obstetricians and Gynaecologists of Canada, 2023).

Another physical symptom of Fetal Cannabis Syndrome is microcephaly, which is a condition in which the head circumference is smaller than normal (Society of Obstetricians and Gynaecologists of Canada, 2023). Microcephaly is associated with a range of neurological problems such as seizures, cerebral palsy, and intellectual disability. Babies with Fetal Cannabis Syndrome may also have facial abnormalities such as small eye openings, flat nasal bridges, and thin upper lips (Society of Obstetricians and Gynaecologists of Canada, 2023).

In addition to physical symptoms, Fetal Cannabis Syndrome is also associated with a range of behavioural and cognitive problems. These problems may persist into childhood and can lead to a range of behavioural problems such as hyperactivity, impulsivity, and inattention (Wu et al., 2011). Cognitive problems associated with Fetal Cannabis Syndrome include difficulties with attention, memory, and language development. Children with Fetal Cannabis Syndrome may have lower IQs and may struggle with academic performance compared to their peers. They may also have difficulties with executive function, which is the set of mental processes that allow individuals to plan, organise, and execute tasks (Wu et al., 2011). While further research needs to be conducted, studies suggest cannabis-related cognitive problems can have long-term consequences on the child's ability to function independently due to these developmental problems.

In addition to physical and cognitive developmental issues, there can be complications for the mother if they use cannabis during the pregnancy term. Infants exposed to cannabis within the uterus may have difficulties with feeding, sleeping, and self-soothing. They may also be more irritable, fussy, and difficult for parents to soothe compared to infants born to non-using mothers (Joseph and Vettraino, 2020). There is also a risk of stillbirth and additional birthing complications during labour and dilation risking both the mother and fetus' health (Joseph and Vettraino, 2020).

Conclusion

As cannabis continues to gain acceptance and become more widely used, there is a need for more research to inform policy decisions and ensure that legalisation is implemented safely and effectively.

One important area for future research is the legal and policy implications of Fetal Cannabis Syndrome. While some states have adopted policies that require healthcare providers to report suspected cases of Fetal Cannabis Syndrome to child protective services, there is limited guidance on how to handle these cases and what support services should be provided to affected families. Future research could investigate the legal and policy implications of Fetal Cannabis Syndrome, as well as identify best practices for addressing Fetal Cannabis Syndrome cases within the healthcare and child welfare systems.

Additionally, there is evidence to suggest that certain groups, such as youth, pregnant women, and people with mental health issues, may be more susceptible to the negative consequences of cannabis use. Future research could investigate the impact of legalisation on these populations, as well as how best to protect them from the potential harms of cannabis use.

Finally, there is a need for more research on the best approaches to cannabis regulation and control. Different countries and states have adopted different approaches to cannabis legalisation, ranging from strict regulation to full commercialisation. Future research could compare and evaluate these different approaches, as well as explore alternative models, such as public ownership or social clubs, to determine which approach is most effective in achieving the goals of legalisation.

In conclusion, as cannabis use becomes more common and accepted, there is a need for more research to inform policy decisions and ensure that pregnant women and their children are protected from the potential harm of cannabis use. Areas of future research could include the prevalence of Fetal Cannabis Syndrome, the impact of cannabis use during pregnancy on maternal and fetal health, the best approaches to preventing Fetal Cannabis Syndrome, the legal and policy implications of Fetal Cannabis Syndrome, and the best approaches to treating and supporting children with Fetal Cannabis Syndrome.

References

American College of Obstetricians and Gynecologists. (n.d.). *Marijuana use during pregnancy and lactation.* ACOG. Retrieved March 3, 2023, from https://www.acog.org/clinical/clinical-guidance/committee-opinion/articles/2017/10/marijuana-use-during-pregnancy-and-lactation

Hammond, D., Goodman, S., Wadsworth, E., Rynard, V., Boudreau, C., & Hall, W. (2020). Evaluating the impacts of cannabis legalisation: The International Cannabis Policy Study. *International Journal of Drug Policy, 77,* 102698. https://doi.org/10.1016/j.drugpo.2020.102698

Joseph P, Vettraino IM. Cannabis in Pregnancy and Lactation - A
Review. Mo Med. 2020 Sep-Oct;117(5):400-405. PMID: 33311738;
PMCID: PMC7723128.

Kavousi, P., Giamo, T., Arnold, G., Alliende, M., Huynh, E., Lea, J.,
Lucine, R., Tillett Miller, A., Webre, A., Yee, A., Champagne-
Zamora, A., & Taylor, K. (2021). What do we know about
opportunities and challenges for localities from cannabis
legalisation? *Review of Policy Research*, *39*(2), 143–169. https://doi.
org/10.1111/ropr.12460

Kroon, E., Kuhns, L., & Cousijn, J. (2021). The short-term and long-
term effects of cannabis on cognition: Recent advances in the field.
Current Opinion in Psychology, *38*, 49–55. https://doi.org/10.1016/j.
copsyc.2020.07.005

Queirolo, R., Rossel, C., Álvarez, E., & Repetto, L. (2019). Why
Uruguay legalised marijuana? the open window of public
insecurity. *Addiction*, *114*(7), 1313–1321. https://doi.org/10.1111/
add.14523

Roncero, C., Valriberas-Herrero, I., Mezzatesta-Gava, M., Villegas,
J. L., Aguilar, L., & Grau-López, L. (2020). Cannabis use during
pregnancy and its relationship with fetal developmental outcomes
and psychiatric disorders. A systematic review. *Reproductive
Health*, *17*(1). https://doi.org/10.1186/s12978-020-0880-9

Society of Obstetricians and Gynecologists of Canada. (n.d.).
Pregnancy info. Learn More about Cannabis and Pregnancy –
Pregnancy Info. Retrieved March 3, 2023, from https://www.
pregnancyinfo.ca/learn-more/#effects

Tattrie, J. (2019). Cannabis Legalisation in Canada. In *The
Canadian Encyclopedia*. Retrieved from https://www.
thecanadianencyclopedia.ca/en/article/marijuana-legalisation-in-
canada

Veldman, M. (2021, September 30). *Cannabis in Uruguay – laws, use,
and more info*. Sensi Seeds. Retrieved March 3, 2023, from https://
sensiseeds.com/en/blog/countries/cannabis-in-uruguay-laws-use-
history/

Wu, C.-S., Jew, C. P., & Lu, H.-C. (2011). Lasting impacts of prenatal
cannabis exposure and the role of endocannabinoids in the
developing brain. *Future Neurology*, *6*(4), 459–480. https://doi.
org/10.2217/fnl.11.27

Chapter 4: Whole Body vs Nervous System

Abdulrahman Aldada

Introduction

A study conducted by the Drug Abuse Department of Health in the USA showed that 4.9% of pregnant women between the ages of 15 and 44 abuse Marijuana. The massive marijuana abuse among young mothers has dramatically led to maternal complications and impairments for newborns (Coliszi & Bhattacharyya, 2020). Moreover, research shows that cannabis use during Pregnancy can cause adverse effects on other body systems, including the cardiovascular and endocrine systems.

The recommendation of further study demonstrates the impact of marijuana usage during gestation on fetal development (Chatlos & Petti, 2021). Due to the increased consumption of Marijuana-infused products, this article reviews the effects of fetal cannabis syndrome on multiple parts of the body, not only the nervous system.

The Endocannabinoid System (ECS) and Pregnancy

The endocannabinoid system is a genetic system of cannabinoid receptors, intrinsic lipid-based retrogressive neuropeptide receptors that connect to cannabinoid nerve endings and cannabinoid binding site peptides that are conveyed throughout the nervous system and external jumpy system of mammals. The biological and cognitive activities regulated by the endocannabinoid system are fertility, gestation, prenatal and postnatal development, hunger, pain perception, mood, and memory. There may be a strong relationship between the effects of cannabis on sexual and immunological functions and the mechanisms that the Endocannabinoid System mediates (Coliszi & Bhattacharyya, 2020). The Endocannabinoid System comprises endogenous cannabis, cannabinoid terminals, lytic enzymes, and membrane hauliers. The endocannabinoid system (ECS) regulates the physiology of Pregnancy, labour, and postnatal development. During Pregnancy, the endocannabinoid affects several functions, including hormone homeostasis and maternal behaviour. Research has shown that the Endocannabinoid System helps control the immune system, reduces inflammation, and protects the developing baby from harm.

In addition, the endocannabinoid receptor may influence the fetus›s cerebral and nervous system growth. Progesterone and estrogen are crucial for the management of endocannabinoid concentrations. Progesterone stimulates lymphocyte fatty acid amide hydrolysis (FAAH)activity via the transcriptional elements and Ikaros, resulting in decreased ethanolamine concentrations. Additionally, research has shown that endocannabinoids regulate the immune cytokine network during reproduction, which seems to be among the fundamental processes regulating implantation and maintaining a healthy pregnancy. Studies on animals have demonstrated the existence of Endocannabinoid System components in the early embryo before neurogenesis, indicating its participation in early embryogenesis (Correa et al., 2018). Thus, maternal Marijuana and synthesised cannabinoid misuse can result in negative reproduction, developmental, and immunological outcomes by significantly changing Endocannabinoid System elements.

Effects of Fetal Cannabis Syndrome on the Brain Development and Cognitive Function

Prenatal cannabis exposure has been associated with the birth of underweight kids, reduced head size, and impaired reflexes in infants. Changes to the growing nervous systems, such as modified neural cell migrations and aberrant synaptogenesis, may be responsible for these infants› impairments. In addition, prenatal cannabis exposure has been linked to cognitive and behavioural issues in offspring. Moreover, prenatal marijuana exposure is associated with reduced IQ scores; infants have reduced attention capabilities and an increased likelihood of behavioural impairments. The disorder is also associated with variations in the growth of the neurological system, including variations in neurotransmitter activity and aberrant brain growth. According to Daha et al. (2020), prenatal cannabis exposure may damage long-term growth and neurological development, especially in cognitive and behavioural development (Daha et al., 2020).

Population-based human research and vitriol animal data suggest that interfering with the endocannabinoid affects normal brain development, specifically synapse maturation, and neuron variability. A longitudinal study of development factors in exposing children to Marijuana and smoking during Pregnancy revealed that cannabinoid children had lower head curvatures at birth, a discrepancy that increases into adolescence. Exposure to Marijuana during the first months of Pregnancy affects head development which correlates with the eventual intelligence quotient.

In contrast, marijuana abuse during the second month of gestation was affiliated with deficiencies in composite, short remembrance, and quantitative scores. These children also exhibit long-term effects, such as visual memory processing capabilities and cognitive issues that might remain until adolescence (Harkany & Cinquina, 2021). Diagnostic magnetic resonance imaging was used to examine the long-term consequences of prenatal cannabis consumption on visuospatial cognition. High levels of mother prenatal cannabis usage were related to considerably increased brain activities in the left inferior and middle frontal gyri, parietal lobe gyrus, middle temporal gyrus, and ganglia of adolescent children between the ages of sixteen to nineteen years.

The reported learning difficulties in these youngsters appear to have a significant impact on their academic performance as early as the sixth grade. The cause of these disorders is unknown. In utero, cannabis exposure disrupts neurotransmitter homeostasis, particularly ventral striatal Dopamine gene modulation and translation. These alterations have increased the likelihood of neuropsychiatric difficulties, such as impulse control difficulties associated with addictive behaviours (Harkany & Cinquina, 2021). In addition, substantial prospective cohort research recorded the consumption of cannabis during Pregnancy by low-income women. However, most research does not support measurable discrepancies in neurodevelopmental consequences in infants aged 0–2 years after cannabis exposure.

However, during the first days in school, cannabinoid kids develop language abilities more slowly than non-cannabinoid children. Infants exposed to Marijuana also exhibit aggressive, violent, and poor affection skills, especially in girls. Cognitive and intellectual deficiencies are also connected to the duration and intensity of prenatal exposure (Diaz Heijtz, 2017). Heavy use of cannabis during the first month of Pregnancy is linked to lower verbal rationale scores in approximately seven hundred six-year-old children compared to their non-exposed peers, found to have a substantially higher risk of developing depressive disorders and attention impairments at the age of ten. In randomised longitudinal research of cannabis-exposed adolescents, the same group of researchers revealed that considerably, the probability of delinquency at 14 years of age was elevated.

Effects of Fetal Cannabis Syndrome on Prenatal and Postnatal Growth

Extracellular cannabinoids have been found to penetrate the fetal–placental barriers in mammals and other animals. The impacts of prenatal, and perinatal cannabis consumption on the initial growth of the fetus and early childhood development of the infant, in addition to the cognition and memory, neurological, and the offspring's social and endocrine characteristics, have been investigated and assessed.

Research has shown that cannabinoids significantly impact immunological response and their impact on the Brain. Although the CB1 receptor is present in the brain and Cannabinoid receptors in distal organs, both receptors are expressed in immune cells (Harkany & Cinquina, 2021). Furthermore, reproductive organs such as uterine endometrial, human Pregnancy, and ovaries have functioning marijuana receptors. However, only a small amount of research has examined its effect on mothers› immune systems and developing fetuses under normal or abnormal situations. These results imply that maternal THC consumption may have persistent effects on offspring immunity.

The survey has shown that Maternal consumption of the psycho-active chemical HU-210 in rodents results in measurable modifications in immune system formation and long-lasting abnormalities to the sympathetic nervous system axis's functioning state. Prenatal treatment of HU-210 decreased the T-helper subgroup in the liver and the proportion of Helper T cells in the circulation of mature male offspring. Additionally, research has shown a rising prevalence of defective lymphocytes in Cannabis-using mothers and their offspring, possibly linking somatic alterations to maternal cannabis usage, which may increase the chance of developing cancer.

According to Kaushal et al. (2019), the effect of short-term cannabis consumption on human placental intestinal epithelial lines discovered that cannabis decreased the activity of placental prostate cancer susceptibility protein. Additionally, cannabis dramatically increased glyburide transport through the human placenta ex vivo, indicating that cannabinoids may increase placental barrier penetration to other xenobiotic compounds and threaten the growing child. An assessment of the self-reported use of illicit Marijuana substances in the women of 538 kids with Marijuana use during the first few weeks of delivery was linked, according to retinoblastoma, with a significantly enhanced risk of retino-blastoma in the offspring, so although its benefits all through late Pregnancy was not linked to a higher risk. In diagnosed cases with neuroblastoma well before the age of one year, the relationship between prenatal cannabis use and cancer incidence was very high (Kaushal et al., 2019). However, this epidemiological research has significant flaws. The data were primarily gathered from hospital interviews, information on the substantial exposure to Marijuana was unavailable, and dose–response analyses were not undertaken; hence, no causal relationship was established.

Effects of Fetal Cannabis Syndrome on Behavioural and Emotional Well-being

Fetal cannabis syndrome (FCS) is a disorder that can afflict children subjected to cannabis during fetal development. There is little study on the psychological well-being of persons afflicted with Fetal Cannabis Syndrome. Existing evidence, however, shows that prenatal marijuana consumption may negatively affect infants' psychological and behavioural development. Children with Fetal Cannabis Syndrome are much more prone to have behavioural and mental issues, including impulsiveness, hyperactivity, anxiety, and anxiety, than children who were not subjected to Marijuana during Pregnancy (Krishnamoorthy & Kaminer, 2020). In particular, prenatal cannabis consumption has been linked to an enhanced risk of social communication difficulties, concentration issues, and poor educational performance.

Surveys also demonstrated a relationship between prenatal cannabis consumption and past elective miscarriages. Women who smoked throughout pregnancy were more likely to report past abortions. Existing statistics on the results of elective abortions for women are disputed. A thorough examination of long-term outcomes for mental health revealed inconclusive indications of psychological suffering following intentional abortion (Wood, 2017). A comprehensive analysis of the evidence on abortion and psychological health, however, indicated a moderate to a significantly elevated risk of mental health issues following abortion: an 81% greater risk of mental health disorders, including drug abuse and suicidal conduct (Koren et al., 2020). Longitudinal research on adolescents revealed an association between induced abortion and mental disorders with an enhanced likelihood of using smoking alcohol, Marijuana, and other illicit substances.

Effects of Fetal Cannabis Syndrome on the Cardiovascular System

Fetal Cannabis Disorder (FCS) is a complicated medical disorder that can affect infants born to parents who smoke cannabis during Pregnancy. Fetal Cannabis Syndrome impacts the physical, emotional, and behavioural health of the infant. Fetal Cannabis

Syndrome may impact several physiological systems, including cardiovascular well-being. According to studies, infants exposed to cannabis throughout Pregnancy could have an elevated risk of cardiovascular illnesses in adulthood. These hazards include high blood pressure, total cholesterol, and reduced risk of stroke (M, 2019). Since Delta-9-tetrahydrocannabinol (THC), the hemp plant's psychotropic constituent, may readily pass the placenta and enter the fetal circulation, it is believed that Fetal Cannabis Syndrome primarily affects the nervous Brain. THC influences the endocannabinoid receptor, which is linked to several crucial brain development processes (Ückert et al., 2018). This can result in abnormalities in brain development, leading to alterations in behavioural, motor, and executive processing and learning capacities. The growing Brain may be more susceptible to these long-term consequences of Fetal Cannabis Syndrome than other sections of the body.

Additionally, infants with Fetal Cannabis Syndrome may be at higher risk for cardiac anomalies, such as ventricular septum and septal abnormalities. Fetal Cannabis Syndrome might result in an irregular heartbeat, manifesting as arrhythmia. Infants with Fetal Cannabis Syndrome may exhibit bradycardia, hypertension, and abnormally fast beats. These heart patterns can raise the risk of cardiopulmonary arrest and other severe cardiac problems (Maia et al., 2020). Moreover, Fetal Cannabis Syndrome may increase the likelihood of obesity and metabolism syndrome, enhancing the possibility of developing cardiovascular disease. Infants with Fetal Cannabis Syndrome must undergo routine check-ups to evaluate their cardiovascular health and ensure that any possible issues are swiftly treated. The renal and urinary system implications of Fetal Cannabis Disorder are still not completely understood. However, evidence indicates that prenatal cannabis consumption may hurt the fetus's kidney and urinary tract growth. Specifically, prenatal cannabis exposure has been linked to a higher prevalence of hydronephrosis, a disorder in which the kidneys expand due to urine buildup. In addition, cannabis consumption during Pregnancy has been related to an elevated risk of renal and urinary tract illnesses and abnormalities of the urinary tract (Member et al., 2020). Lastly, prenatal marijuana consumption may alter renal circulation flow, resulting in long-term kidney issues.

According to Mazza et al. (2021), Adolescent mothers are more likely to have Fetal Cannabis Syndrome because they are more likely to consume cannabis during Pregnancy. The increased marijuana abuse among young mothers may be attributed to them being unaware of the possible hazards of cannabis usage during Pregnancy. Additionally, young moms may have limited access to healthcare and might not receive adequate knowledge about the dangers of cannabis use during Pregnancy (Mazza, 2021). Consequently, Fetal Cannabis Disorder is more prevalent among young mothers. The most effective strategy to prevent Fetal Cannabis Syndrome is to abstain from cannabis usage throughout Pregnancy. If a pregnant woman chooses to use cannabis, they should proceed with the utmost caution, as it is still unknown how much cannabis may be used safely during Pregnancy. In addition, pregnant women should consult their healthcare physician before taking cannabis. Finally, pregnant women should be conscious of the potential hazards linked with cannabis use and communicate any issues with their doctor.

Effects of Fetal Cannabis Syndrome on the Immune System

Cannabis usage among women of reproductive age appears to have increased. In one National Institute of Drug Abuse tracking the Future survey, 10.4% of nineteen- to thirty-two-year-old women acknowledged smoking cannabis. A recent poll found that roughly 4.7% of women of reproductive age consume cannabis regularly. The incidence of drug addiction during Pregnancy varies between 5% and 16%. Five million women of reproductive age are believed to use illegal drugs (Mazza, 2021). An estimated 500,000 newborns are exposed to one or more illegal chemicals in the uterine in the United States. As a result, maternal drug addiction harms the mother and her children, creating a serious public health issue. According to the National Pregnancy and Wellness Survey performed by the National Research centre on Substance Abuse, the predominance and content use trends among women trying to produce live-born newborn babies. Self-reported weed use during Pregnancy in America was 2.9%, compared to opioid use, which ranged from 0.9% to 1.1% (Mazza, 2021). While the percentage of pregnant women admitted to drug misuse treatment in the United States stayed at 4% between 1992 and 2012, the proportion of pregnant women who indicated cannabis usage climbed significantly from 29 to 43% (Mazza, 2021).

Research has shown that Pregnant women who abuse other illegal substances are more prone to using cannabis than the ones who do not abuse any drugs. The increased substance abuse is commonly attributed to the belief that cannabis is less dangerous to the growing embryo and fetus than other substances such as morphine, opium, or amphetamine. Cannabis use among women also during delivery is expected to increase in the coming years due to the licensing and legalisation of Marijuana for medicinal and hedonistic purposes in several states. According to one potential mechanism, Marijuana may interfere with the development of the embryonic immunological system and reduce the body›s ability to fight illness (Papaioannou et al., 2022). It has been demonstrated that using cannabis while pregnant increases the risk of having a baby with a lower birth weight, a premature birth, and a weakened immune system. Additionally, Fetal Cannabis Syndrome has been associated with an increased risk of acquiring allergies and asthma. This may result from a poor immunological response, raising the risk of asthma and allergy due to a failure to react appropriately to antigens or irritants.

Additionally, prenatal cannabis consumption has been linked to an increased chance of acquiring autoimmune illnesses, such as inflammatory arthritis and type-1 diabetes, a condition brought on by the body›s immunological system destroying its tissues and organs. This may be attributed to a compromised immune system, which may increase the likelihood of acquiring autoimmune illnesses. Fetal Cannabis Syndrome may substantially influence the immune response and raise the possibility of developing various health issues (Pimentel et al., 2021). To limit the danger of health issues in the baby, pregnant women must abstain from cannabis usage.

During Pregnancy, the mother›s immune response actively endures the semi-allogeneic baby. This encompasses abnormalities in the endometrial mucosa›s local immune function and peripheral immunological reactions. During gestation, the immune system is stimulated. During a full-term pregnancy, granular monocytic lineage cells grow and experience phenotypic and operational stimulation while dendritic cell counts decline. Additionally, the frequency of native killer T lymphocytes and the generation of interferon (IFN)- by NK cells are reduced in pregnant women. Certain infections and immunological dysregulation produced by proinflammatory or suppressive stressors are hazardous for pregnant women. This means that maintaining a healthy fetus during Pregnancy and ensuring effective placentation require a drastically altered immune system (Saito et al., 2021). However, interference with the body›s immune response during Pregnancy might disrupt the equilibrium between resistance and susceptibility and affect the result.

Conclusion

Research has shown that the whole body is affected by Fetal Cannabis Syndrome, not only the neural system. In addition to the cardiovascular and hormonal systems, studies show that cannabis use during Pregnancy may harm other bodily systems, such as the immune system. However, there is a need for more studies to completely comprehend the various consequences of cannabis use during gestation on embryonic development, as well as other parts of the body of developing infants. After detailed research, all the possible hazards should be documented as a precaution for all pregnant mothers, especially adolescent mothers. Expectant mothers should talk with their health professionals about the potential dangers of marijuana usage and address any concerns (Shah, 2019). The healthcare professionals will give guidelines on how to quit the marijuana addiction and various measures to ensure that the developing fetus is safe. The endocannabinoid system (ECS) regulates the physiology of Pregnancy, labour, and postnatal development. During Pregnancy, the Endocannabinoid System affects several functions, including hormone homeostasis and maternal behaviour. Marijuana misuse can result in negative reproduction, developmental, and immunological outcomes. Pregnant women who abuse other illegal substances are less prone to use cannabis. Marijuana may interact with the proper growth of the embryonic immune system and diminish the body›s capacity to battle sickness. Cannabis usage during Pregnancy increases the chance of lower birth weight, premature delivery, and a weaker immune system in the infant.

References

Adamson, M., Di Giovanni, B., & Delgado, D. H. (2020). The positive and negative cardiovascular effects of cannabis. *Expert Review of Cardiovascular Therapy*, 1–13. https://doi.org/10.1080/14779072.2020.1837625

Barbosa-Leiker, C., Burduli, E., Smith, C. L., Brooks, O., Orr, M., & Gartstein, M. (2020). Daily Cannabis Use During Pregnancy and Postpartum in a State with Legalised Recreational Cannabis. *Journal of Addiction Medicine*, 1. https://doi.org/10.1097/adm.0000000000000625

Chatlos, J. C., & Petti, T. A. (2021). Implications of Cannabis Legalisation: A Medical Perspective. *Adolescent Psychiatry*, *11*. https://doi.org/10.2174/2210676611666210616144139

Coliszi, M., & Bhattacharyya, S. (2020). Cannabis: Neuropsychiatry and Its Effects on Brain and Behaviour. *Brain Sciences*, *10*(11), 834. https://doi.org/10.3390/brainsci10110834

Daha, S. K., Sharma, P., Sah, P. K., Karn, A., Poudel, A., & Pokhrel, B. (2020). Effects of prenatal cannabis use on fetal and neonatal development and its association with neuropsychiatric disorders: A systematic review. *Neurology, Psychiatry, and Brain Research*, *38*, 20–26. https://doi.org/10.1016/j.npbr.2020.08.008

Douglas, J., Nelson, D. M., Sunga, P., & Smith, D. (2020). Cannabis Use and Baby Boomers: Attitudes and Patterns of Consumption. *Cannabis*, *3*(1), 1–10. https://doi.org/10.26828/cannabis.2020.01.001

Gonzalez-Mota, A., Covacho-Gonzalez, M., Valriberas-Herrero, I., Roncero, C., & De La Iglesia-Larrad, J. (2021). A systematic review about the screening of cannabis use during pregnancy and neonates. *European Psychiatry, 64*(S1), S825–S825. https://doi.org/10.1192/j.eurpsy.2021.2180

Harkany, T., & Cinquina, V. (2021). Physiological Rules of Endocannabinoid Action During Fetal and Neonatal Brain Development. *Cannabis and Cannabinoid Research.* https://doi.org/10.1089/can.2021.0096

Koren, G., Cohen, R., & Sachs, O. (2020). Use of Cannabis in Fetal Alcohol Spectrum Disorder. *Cannabis and Cannabinoid Research.* https://doi.org/10.1089/can.2019.0056

Krishnamoorthy, D., & Kaminer, Y. (2020). The Effects of Prenatal Exposure to Marijuana on Early Childhood Development: A Systematic Literature Review. *Cannabis, 3*(1), 11–18. https://doi.org/10.26828/cannabis.2020.01.002

Mahabir, V. K., Merchant, J. J., Smith, C., & Garibaldi, A. (2020). Medical cannabis use in the United States: a retrospective database study. *Journal of Cannabis Research, 2*(1). https://doi.org/10.1186/s42238-020-00038-w

Maia, J., Fonseca, B., Teixeira, N., & Correia-da-Silva, G. (2020). The fundamental role of the endocannabinoid system in endometrium and placenta: implications in pathophysiological aspects of uterine and pregnancy disorders. *Human Reproduction Update, 26*(4), 586–602. https://doi.org/10.1093/humupd/dmaa005

Mamber, S. W., Gurel, V., Lins, J., Ferri, F., Beseme, S., & McMichael, J. (2020). Effects of cannabis oil extract on immune response gene expression in small human airway epithelial cells (HSAEpC): implications for chronic obstructive pulmonary disease (COPD). *Journal of Cannabis Research, 2*(1). https://doi.org/10.1186/s42238-019-0014-9

Mazza, M. (2021). Medical cannabis for treating fibromyalgia syndrome: a retrospective, open-label case series. *Journal of Cannabis Research, 3*(1). https://doi.org/10.1186/s42238-021-00060-6

Reece, A. S., & Hulse, G. K. (2020). Canadian Cannabis Consumption and Patterns of Congenital Anomalies: An Ecological Geospatial

Analysis. *Journal of Addiction Medicine, 14*(5), e195–e210. https://doi.org/10.1097/adm.0000000000000638

Reece, A. S., & Hulse, G. K. (2022). European epidemiological patterns of cannabis- and substance-related congenital cardiovascular anomalies: spatiotemporal and causal inferential study. *Environmental Epigenetics, 8*(1). https://doi.org/10.1093/eep/dvac015

Roncero, C., Valriberas-Herrero, I., Mezzatesta-Gava, M., Villegas, J. L., Aguilar, L., & Grau-López, L. (2020). Cannabis use during pregnancy and its relationship with fetal developmental outcomes and psychiatric disorders—a systematic review. *Reproductive Health, 17*(1). https://doi.org/10.1186/s12978-020-0880-9

Shah, N. (2019). Joubert Syndrome: Two Different Prenatal Ultrasound Presentations. *Journal of Fetal Medicine, 6*(1), 41–43. https://doi.org/10.1007/s40556-019-00195-w

Ückert, S., la Croce, G., Bettiga, A., Bannowsky, A., Przigoda, L., Kuczyk, M. A., & Hedlund, P. (2018). 305 Expression and Distribution of Key Proteins of the Endocannabinoid System (ECS) in the Human Seminal Vesicles. *The Journal of Sexual Medicine, 15*(2), S78. https://doi.org/10.1016/j.jsxm.2017.11.188

Wood, D. (2017). The cardiovascular effects of cannabis and cannabinoids: Myth or reality? *Toxicology Letters, 280*, S71. https://doi.org/10.1016/j.toxlet.2017.07.182

Chapter 5: Prevention -Investigating whether there is any way to Prevent Fetal Cannabis Syndrome and the Importance of not using Cannabis During Pregnancy

Annabella Stoll-Dansereau

Introduction

As my colleagues have highlighted in chapters prior it is apparent the severity of cannabis syndrome is. It has been shown that when mothers ingest and interact with cannabis, they increase the risks associated with their children being born with significant challenges relative to their peers. A key reason cannabis can be so damaging is it impacts the brain in the crucial gestation period when in the mother's uterus. Not only does this impact the babies through things like reduced dopamine levels it has lasting impacts on children and potentially adults. For example, one study conducted as a time series showed results of children being worse off on many measures including fine motor skills and raw intelligence scores (Wu et al., 2011). This has statistically significant negative effects on people that are exposed to cannabis repeatedly as a prenatal. However, this entire disease like impairments put onto children is entirely preventable. There is nothing wrong with the children's DNA yet some of the symptoms mimic genetic issues or mental health challenges (Baranger et al., 2022). The positive is that this is entirely preventable. Should the mother not use cannabis during pregnancy the child would never face the struggles caused by this disease. Therefore, it is integral that as cannabis is more mainstream and legalised, we must help prevent this syndrome from impacting our young. Although this is much easier said than done. Many people use cannabis for a reason. One odd party in which cannabis is consumed may not have an impact a prenatal development. It is the repeated exposure that is the biggest issue.

This repetition can come from many avenues although two I will highlight are the necessity for physical pain and emotional pain. These can lead to addiction, and I will touch more on the dangers of the addiction cycle. Hopefully, this will give the reader a better understanding and empathy for why someone may still be using cannabis while pregnant. Another reason people may be using has less to do with addiction and more to do with a lack of education. Some may not be aware of the implications to the children of their continued use as such this will be a keystone in the latter half of this paper discussing prevention methods.

Why Mothers may be using Cannabis: Physical Pain

Medical cannabis has a long history of providing natural alternatives to manage and help with chronic pain or ongoing diseases. Before cannabis was legal for the general population its medicinal properties were recognised as such that depending on your situation you could get legal access to cannabis. While there is no distinction between users today this shows the history of using this plant for medical reasons. Some of these which it is used in such as cancer do not apply to pregnant women as they most likely would not be battling cancer while getting chemotherapy. However, some of the diseases that can be treated with cannabis are much more common and chronic issues that do not completely obstruct one's life.

The most obvious use is for general chronic pain be this from a disease or stress with a laborious job such as a faulty back as a construction worker. In a study by Whiting et al. (2015), it was shown that perceived pain was reduced by around 40% compared to the control group. So pregnant women that may be dealing with pain daily might rely on this substance before they know they were carrying a child. Another use of cannabis is for Tourette's. No studies have proved this to be an effective cure or reducer of tics and symptoms associated with Tourette's however on an anecdotal level many people with this disease swear that this reduces compulsive movements. So even if it is not a proven medicine even if a large subsection of a group uses it, we must consider this when devising a prevention plan. The final point on health problems addressed by cannabis is traumatic brain injuries such as a concussion. It was shown by Di Napli et al. (2016) that outcomes for those with brain injuries improved when they consumed cannabis. While short-term concussions might not result in someone needing to take cannabis on a cyclical basis many brain injuries drag on leaving someone suffering for many years after the post-traumatic brain injury.

None of these health reasons for using cannabis holds a significant number of pregnant people however taking all the medical reasons one uses cannabis the chances of people being pregnant while taking cannabis increases. Ways to prevent this type of addiction will be addressed below. The most troubling form of cyclical cannabis use is for psychological pain or mental health struggles. This subset most likely wouldn>t have qualified for a medical cannabis licence but find themselves needing cannabis to cope with the daily struggles of life and living inside their head.

Why Mothers may be using Cannabis: Psychological Pain

Common effects of cannabis use are relaxation and an overarching feeling of happiness. In scientific terms can have "anxiolytic and euphoriant effects" (Grunberg et al., 2015). You might infer that this is extremely valuable for those who feel out of control, depressed or anxious. To be clear there is no link that people feeling down actually improve their mental challenges through the use of cannabis there have been studies suggesting that cannabis use may increase the risk of mental health diseases such as depression and even schizophrenia (Hall et al., 2008). However, even in light of this many people will use cannabis as an unhealthy coping mechanism. This must be recognised when determining prevention plans. It is essential to have compassion for people's pain when telling them they must stop this action until they have taken their baby to term.

These two reasons, physical and psychological pain both cause addiction to cannabis. While it may be generally thought of as a recreational drug with less harmful addiction properties than typical problem drugs such as heroin, we still see the dangerous cycles of addiction.

How Addiction to Cannabis Works

Originally cannabis was thought to be a dangerous and addictive substance however more research is coming out showing the side effects of this drug. As with many hard drugs cannabis use causes an increase in dopamine release. Historically this effect was documented in animal trials (Ng Cheong Ton et al., 1988) where they administered TCH and saw a rise in this chemical. Additionally, we have seen more current trials showing this pertains not only to animals but also humans (Stokes et al. 2010). We also see that recurring cannabis users have a higher risk of becoming depending and establishing a substance use disorder. One study found that almost ten per cent of people using cannabis would qualify for this dependence diagnosis on the DSM-IV (Volkow et al., 2014). Therefore, we cannot discount the harm this substance creates even when it is legal.

As with many addictions cannabis still causes the three stages of preoccupation/anticipation leading to binge intoxication then withdrawal effects leading back into the first part of the cycle. In the preoccupation stage, we see possible impairments to memory and IQ because of impaired executive function and disrupted glutamate signalling which s correlated with relapse and craving. If one gives into the cravings, they will become impaired by the drug with changes in their reward processing as dopamine is released. Finally, after completing the drug high they will lead themselves into the withdrawal stage where one can adopt a negative effect associated with the dysregulation of the amygdala (Zehra et al., 2018)

As you might imagine future and current mothers are still subject to this addiction cycle like any other person. While one might know not to try cannabis and become addicted while pregnant many may already be battling this cycle before pregnancy and be unable to break free.

Education for Proactive Prevention

A natural question to ask is whether the general population understands the risks associated with this drug. Cannabis is typically thought of as a *safe drug* although this is a bad mentality. In the same way, alcohol is legal it doesn›t mean it is good to be dependent on it. Like anything cannabis isn›t inherently bad and provides many tangible benefits such as a reduction in pain described above but this must not bleed into addiction territory. So there need to be two modularities of education one for the general population and another for known mothers.

In terms of the general population, we need to highlight the dangerous addictive properties of this drug- in the same way, that cigarette packages were mandated to show the terrible side effects that smoking has the government may want to think about bringing these principles to cannabis products. It was proposed by Orenstein et. al that governmental bodies must step in and create standardised packaging abiding by a few key principles. First, the packaging must be plain and simple with clear health warnings. Secondly, they suggested governments ban flavoured products and any addictive additives some products contain. This could help shift the perception letting people see these drugs for what they truly are, addictive substances (2018).

This relates to pregnant women because pregnant woman comes from the general population so if they are already addicted before pregnancy the road to detox is much more difficult. So, if people understood the harmful addictive properties of this drug, they might think twice about starting to use this drug especially if future pregnancy is a goal.

The second way of education would be educating actual mothers. Most mothers will see a doctor at various points throughout their pregnancy thus doctors should bring it to all mothers' attention in the same way alcohol and pregnancy are mentioned. Additionally, educational pamphlets could be beneficial in laying out the harmful effects this drug has on a fetus. Now clearly this education will not solve the issue should one be already addicted or using it for pain however it will increase the general awareness of this issue and could help prevent this from occurring in future generations. The rest of this chapter will address how we may prevent this harmful syndrome in vulnerable populations where mothers use the drug already.

Prevention: Learning for Fetal Alcohol Prevention

As highlighted above the doctors should bring this issue to the attention of the mothers. Most will just agree and not have an issue however some mothers may already be addicted. Physicians can use a similar approach to fetal alcohol intervention techniques to identify these individuals. First, the doctor should ask all women who use or have used cannabis a questionary similar to the Alcohol Use Disorders Identification Test-Consumption. Then depending on the risk level identified by the test we should refer the pregnant women to the appropriate services. For example, in this alcohol pregnancy intervention study people would score a medium risk were referred to 'get healthy in pregnancy telephone coaching services and those who were high risk were referred to 'drug and alcohol clinical services. In giving these support outlets to heavier drinkers this study found an improvement in outcomes however in the less at-risk individuals just being told drinking is bad didn't statistically help anything.

Prevention: For Pregnant Cannabis Users

So here we get to an issue about resource allocation. But for a moment we will put this on hold and describe what would happen in a perfect work anyone that was determined to be at risk by the said test would be led to specialised healthcare providers that could provide support.

In our earlier section, it was highlighted that dependence may originate from the alleviation of pain. In this case, doctors should work with individuals to find pain management alternatives. This varies widely from individual to individual depending on the source of the pain. So, there should be phone lines similar to the nurse hotline where pregnant women can receive advice about alternative ways to manage their pain which will not impact the baby. This dependence is simpler should the women not be fully addicted to the substance and just use the substance to help improve their quality of life.

Some people may have resorted to using cannabis to help low levels of pain such as back pain without seriously looking into alternative medications because cannabis is legal and easily accessible today. While no studies have been done to address this hypothesis, it may be the case that if women can easily call someone to find out alternatives to address their pain, they could both live the quality of life they enjoy with cannabis while still producing a strong healthy baby. This number should be given out to everyone regardless of it they were screened for at-risk by the doctors and included on the pamphlet given out to pregnant individuals. We need to provide alternatives to doctor websites which many people can get their information from today. In speaking with a qualified healthcare professional in an anonymous online context the pregnant woman may feel less shame and feel safer to fully disclose why she uses the drug without fear of judgment. And the nurse or doctor would be able to continue to educate on the harm you are putting your child in when you continue to use this substance. Should one be unable to leave the drug because of a physical ailment they may not be in the position to have a child and alternatives to pregnancy should be addressed by these practitioners.

For the second reason, addiction, depending on how deeply ingrained the cycle has become different tiers of support should be available from counselling sessions to help people navigate the path away from this endless loop to in-person rehab programs. In the test described above to screen dependence on alcohol one like that for cannabis could let the healthcare system know what level of support and care is required.

If a pregnant woman poses a moderate risk of bringing a baby into this world with fetal cannabis syndrome, they should be connected with a counsellor to provide a predetermined number of sessions sponsored by the government. In these sessions, practitioners could work on alternatives to drugs to feel better about life challenges. A key element to these sessions would be putting in support systems to help the mother should she feel the craving and desire to get a high from the drug. With these sessions, a mother has someone to be accountable to if she cannot be accountable for the fetus growing inside of her.

After the questionnaire, if the woman scored severely high risk, then it would be recommended, she goes into a rehab facility again hopefully sponsored by the government. This would provide a safe environment away from the drug so that the woman could get out of the depths of addiction. If someone scored so high on this quiz it is probably the case that they cannot get out of this addiction and even the motivation of hurting the baby won't be able to override the mental anguish they feel when trying to go against the addiction.

Regardless of the severity sufferers that use this drug for the *happy feeling* and rely on it emotionally need to learn new techniques to deal with their emotions such that they don't feel like they need this drug.

This article by Roncero et al. perfectly summed up what was described above where it was stated how "Healthcare professionals play an essential role both in psychoeducational counselling, awareness and prevention initiatives and in the early detection of cannabis use and the diagnosis and treatment of cannabis abuse and dependence" (2020). So, it is clear services are needed however the healthcare system is always tight on resources and funding making the required support in an ideal world not necessarily achievable. From an economic standpoint, resources flow where money is invested. So, the government needs to take a proactive stance in addressing this syndrome after legalising this drug. They should invest in educational programs for the general population and targeted support programs for those that are both pregnant and addicted to try and target the root cause of the issue.

Conclusion

At the beginning of this chapter, we discussed the harms of using cannabis while pregnant and the other chapters of this book will address these harms in greater detail. After that, we addressed why one might turn to cannabis and how this substance has addictive properties. Finally, we looked at potential prevention methods addressing the key reasons established above surrounding why one might be using cannabis. A key finding was that we need more healthcare and mental health professionals to provide these services funded by the government.

Another finding of this chapter was how the general population needs to be educated. At first glance, it may seem like only pregnant women need to be aware of this issue but that is far from the case. When we look at alcohol and pregnancy it is extremely apparent this connection and almost everyone in society has become aware of this from ads to pamphlets, they have seen over the years in healthcare centres and liquor stores. Therefore, it has become not socially acceptable to drink while pregnant because everyone around you is aware that you are harming your future baby. The same general education should be applied to cannabis. While it is very important to provide support to actual pregnant women in the ways described above, we also want to make sure people don't make a mistake getting pregnant and still using cannabis. Additionally, you would no longer have the social aspect because going to a dispensary or smoking with your friends would become a judged experience as the general public was aware of the harm you were subjecting your child to. Finally, in general, population education women that were dependent on cannabis may choose to not get pregnant in the first place if they know they couldn't get off this drug and knew the harm it would cause their baby.

Ultimately, this syndrome is going to be hard to address since no education or service is surely going to solve the problem but we must start acting regardless before more babies are brought into this world with a preventable syndrome. Many of these preventions require governmental or alternative non-profit funding however this funding is necessary to protect the future generation from Fetal Cannabis Syndrome.

References

David A. A. Baranger. (2022, December 1). *Association of Mental Health Burden with prenatal cannabis exposure from childhood to early adolescence.* JAMA Pediatrics. Retrieved March 3, 2023, from https://jamanetwork.com/journals/jamapediatrics/article-abstract/2795863

Di Napoli M, Zha AM, Godoy DA, Masotti L, Schreuder FH, Popa-Wagner A, Behrouz R. Prior cannabis use is associated with outcome after intracerebral hemorrhage. Cerebrovascular Disease. 2016;41(5-6):248–255.

Grunberg, V. A., Cordova, K. A., Bidwell, L. C., & Ito, T. A. (2015, September). *Can marijuana make it better? prospective effects of marijuana and temperament on risk for anxiety and depression.* Psychology of addictive behaviours: journal of the Society of Psychologists in Addictive Behaviours. Retrieved March 3, 2023, from https://www.ncbi.nlm.nih.gov/pmc/articles/PMC4588070/

Hall, W., & Degenhardt, L. (2008). *Cannabis use and the risk of developing a psychotic disorder.* World psychiatry: official journal of the World Psychiatric Association (WPA). Retrieved March 3, 2023, from https://www.ncbi.nlm.nih.gov/pmc/articles/PMC2424288/

Koppel BS, Brust JC, Fife T, Bronstein J, Youssof S, Gronseth G, Gloss D. Systematic review: Efficacy and safety of medical marijuana in selected neurologic disorders: Report of the Guideline Development Subcommittee of the American Academy of Neurology. Neurology. 2014;82(17):1556–1563.

Ng Cheong Ton, J. M., Friedemann, G. A., & Etgen, A. M. (1988). *The effects of Delta 9-tetrahydrocannabinol on potassium-evoked release of dopamine in the rat caudate nucleus: An in vivo electrochemical and in vivo microdialysis study.* Brain research. Retrieved March 1, 2023, from https://pubmed.ncbi.nlm.nih.gov/2855215/

Roncero, C., Valriberas-Herrero, I., Mezzatesta-Gava, M., Villegas, J. L., Aguilar, L., & Grau-López, L. (2020, February 17). *Cannabis use during pregnancy and its relationship with fetal developmental outcomes and psychiatric disorders. A systematic review -*

reproductive health. BioMed Central. Retrieved March 1, 2023, from https://reproductive-health-journal.biomedcentral.com/articles/10.1186/s12978-020-0880-9

Stokes, P., Egerton, A., & Watson, B. (2010, May 6). *Significant decreases in frontal and temporal [11C]-raclopride binding after THC Challenge.* NeuroImage. Retrieved March 3, 2023, from https://www.sciencedirect.com/science/article/pii/S1053811910007020?casa_token=WYIiDaceqw8AAAAA%3Ah-6Pa5004cG6m6grZ5_KOZ9EPzqEjOasgBy2icBuVIYelxxXo-QEq1Iz4omzDSxhr3l0cjZSlh

Volkow, N. D., Baler, R. D., Compton, W. M., & Weiss, S. R. B. (2014, June 5). *Adverse health effects of marijuana use.* The New England journal of medicine. Retrieved March 3, 2023, from https://www.ncbi.nlm.nih.gov/pmc/articles/PMC4827335/

Whiting PF, Wolff RF, Deshpande S, Di Nisio M, Duffy S, Hernandez AV, Keurentjes JC, Lang S, Misso K, Ryder S, Schmidlkofer S, Westwood M, Kleijnen J. Cannabinoids for medical use: A systematic review and meta-analysis. Journal of the American Medical Association. 2015;313(24):2456–2473.

Wu, C.-S., Jew, C. P., & Lu, H.-C. (2011, July 1). *Lasting impacts of prenatal cannabis exposure and the role of endogenous cannabinoids in the developing brain.* Future neurology. Retrieved March 3, 2023, from https://www.ncbi.nlm.nih.gov/pmc/articles/PMC3252200/

Zehra, A., Burns, J., Liu, C. K., Manza, P., Wiers, C. E., Volkow, N. D., & Wang, G.-J. (2018, December). *Cannabis addiction and the brain: A Review.* Journal of neuroimmune pharmacology: the official journal of the Society on NeuroImmune Pharmacology. Retrieved March 1, 2023, from https://www.ncbi.nlm.nih.gov/pmc/articles/PMC6223748/

Chapter 6: COVID and Increased Cannabis Use in Canada – Discussing the Impact of Increased Cannabis use during the COVID Pandemic and its Potential Effects on Fetal Development

Holly Steen

Introduction

The COVID-19 pandemic saw increased rates of anxiety, depression, and substance abuse among the adult population in Canada. Public health restrictions focused on quarantine and social isolation, while useful in reducing the spread of the virus, have been linked to increased levels of loneliness and social isolation, resulting in repercussions for physical and mental health (Hwang et al., 2020). A 2020 study found that 50% of Canadians reported a decline in their mental health during the pandemic (Angus Ried Institute, 2020). Substance use also increased during the COVID-19 pandemic, with those reporting poorer levels of mental health more likely to consume alcohol, cannabis and tobacco (Centre for Addiction and Mental Health, 2020).

This chapter will explore the possible impacts of increased cannabis use during the COVID-19 pandemic on fetal development in Canada. It will look at the time frame from January 2020, when the first COVID-19 case was reported in Canada (Canadian Institute for Health Information, 2021), to the end of 2022. It will consider factors that may have influenced women's decisions to use cannabis during pregnancy, specifically the social acceptance of using cannabis, the impact of COVID-19 isolation procedures on cannabis use, and access to healthcare resources. The second half will then explore the potential effects of that use on fetal development.

Social Acceptance & Use of Cannabis

The legalisation of marijuana in 2018 has impacted the social acceptability of the substance in Canada. The 2020 Canadian Cannabis Survey reported that among alcohol, tobacco and cannabis, cannabis is the second most socially acceptable substance. In both 2019 and 2018, the study shows a slight year-by-year increase in the social acceptability of occasionally using cannabis (Health Canada, 2019, 2020). Social acceptability is linked to cannabis use. When compared with data from 2018 and 2019, it can be observed that cannabis use has increased. A study found that by the end of 2020, 20% of Canadians aged 15 or older reported having used cannabis in the past three months, which was higher compared to reported rates of 14% before legalisation in October 2018 and 17.5% in the months following the enactment of the *Cannabis Act* (Rotermann, 2021). Importantly, higher self-reported use of cannabis use since the legalisation of Canada is not only impacted by the upward trend in cannabis consumption but an increased willingness to report cannabis use due to legalisation removing stigma and illicitness around the drug (Koto et al., 2022). Beyond increased cannabis use by the general public, it has been observed that increased public acceptance and support of marijuana use and its legalisation are likely to increase the prevalence of perinatal marijuana use (Holland et al., 2016; Koto et al., 2022).

COVID-19 circumstances & cannabis use

Many factors can influence one's decision to consume cannabis while pregnant. Socioeconomic and demographic factors associated with maternal cannabis use include younger maternal age, low education and low income (Brown et al., 2019; Koto et al., 2022). Cannabis may be used for the management of pregnancy symptoms, psychical and mental, along with other health issues (Nashed et al., 2021). Many of the stressors specific to the pandemic or which increased during this time are associated with increased substance use, which includes cannabis use. Pandemic-specific stressors include isolation restrictions, experiencing pandemic-related financial difficulties and food insecurity or dietary changes.

Isolation requirements and concerns about COVID-19 have negatively impacted Canadians' mental health. In a report on COVID-19's impact on harms due to substance use, the Canadian Institute for Health Information observed that the pandemic has had an unprecedented disruption on the lives of Canadians (2021). This includes pregnant women. During the COVID-19 pandemic, certain symptoms, specifically those related to mental health, were linked with increased cannabis use in pregnant women (Kar et al., 2021; Nashed et al., 2021). A 5-year Canadian study from February 2009 to February 2014 examined mental health risks for drug use during pregnancy and found that depression was the top risk factor for the use of alcohol, tobacco and cannabis during pregnancy, with women who were depressed during pregnancy being 2.56 times more likely to use cannabis (Brown et al., 2019).

Self-medicating depression and anxiety through cannabis use while pregnant is a growing medical concern. One of the core components of cannabis, CBD, is commonly perceived to reduce symptoms of pain, anxiety and depression during pregnancy, though the safety of CBD use in pregnancy is unknown (Nashed et al., 2021). The prevalence of mood and anxiety disorders in pregnant women is higher than in the general population, as "pregnancy is a vulnerable developmental period when expectant mothers can experience recurrence of mental disorders" (Eleftheriades et al., 2022). The use of substances to self-medicate or as a coping mechanism can be a result of new or worsened symptoms of depression during the pandemic (Kar et al., 2021).

A 2021 study out of the University of Calgary on alcohol and substance use in pregnancy during the COVID-19 pandemic with 7470 pregnant participants, found that 12.8% of pregnant individuals reported using at least one substance after learning they were pregnant, with 4.3% being cannabis use. This was contrasted by the year before their pregnancy, when 66% of pregnant individuals reported having used at least one substance, including alcohol, with 17.9% reporting cannabis use. There was a significant link between mental health and substance use, with researchers finding a strong association between higher levels of cannabis and tobacco in pregnant women and higher depression symptoms. Several other pandemic-related factors were associated with more cannabis use in participants, including feeling worried about COVID-19 threatening their baby's life, worrying about not receiving care for themselves and their baby, and feeling socially isolated (Kar et al., 2021).

Pregnancy is already a vulnerable time for pregnant women financially due to the additional costs of prenatal care and acquiring newborn necessities (Shirreff et al., 2021). Pregnant women experiencing COVID-19-related financial difficulties were associated with higher reported use of cannabis, and tobacco, and co-using more substances (Kar et al., 2021). This is of concern since "more than 40% [of pregnant women] reported a decrease in their income due to the COVID-19 pandemic" (Vagher-Mehrabanii et al., 2022).

Access to healthcare resources and information

With pregnant women facing higher rates of mental health conditions during the pandemic and observed increases in cannabis use, access to clear information on cannabis use while pregnant was an important factor in providing individualised perinatal care. Though, barriers to accessing this information have drawn further attention to the need for physician education on perinatal cannabis use and how to counsel patients on the topic.

With pregnancy being a critical time for fetal development, prenatal care providers play the important role in educating expecting women on healthy lifestyle choices. Though, before COVID-19 first entered Canada, a lack of communication between physicians and pregnant patients on the dangers of cannabis use during pregnancy was already observed. A 2019 Alberta study found that among the topics of nutrition, weight management and substance abuse prevention, the last was the least consistently discussed topic among perinatal care providers including midwives, family physicians and obstetricians (Premji et al., 2019).

A 2019 study out of Hamilton, Ontario examined 478 pregnant women's perceptions regarding cannabis use while pregnant. The study asked participants where they would look for information if they wanted to learn more about cannabis. It was founded that the internet was the most likely to be consulted (75.1%), followed by healthcare professionals (66.9%). Though, seeking information from the internet, along with other less-reputable sources, such as pregnancy groups, family or friends and cannabis dispensaries, risks the potential for accepting misinformation to reconcile negative feelings they may have around cannabis use while pregnant. Women who did not believe cannabis could be transmitted from mother to fetus were more likely to ask less-reputable sources. Among all these sources, healthcare providers are in the best position to not only provide education on cannabis use but access to further resources and an approach developed according to patients' individual needs. This may include a harm reduction approach in cases when patients are unable or unwilling to stop cannabis use while pregnant (Bartlett et al., 2020).

Healthcare providers' ability to speak to pregnant patients about marijuana use requires that these providers be educated on research regarding the long-term effects and outcomes of marijuana use during pregnancy. Cannabis legalisation and its likelihood to increase perinatal cannabis use call for new protocols and standardised education for physicians to follow when counselling patients on the use of cannabis while pregnant. The legal approach to discouraging cannabis, which physicians could once default to before the legalisation of cannabis, is no longer an option. Part of addressing physicians' lack of education on the topic calls for modification of curriculums and residency training to expose doctors to this knowledge (Holland et al., 2016). Another important factor is research gaps - researchers have consistently noted the need for further research on perinatal cannabis use and its potential impacts on the fetus. This is in part due to past research standards and studies which do not isolate cannabis use with other substance use among other limitations (Nashed et. al, 2021; Vanstone et al., 2022). It is not simply more extensive research needed but definitive research that healthcare providers can apply when counselling pregnant women about perinatal cannabis use (Bartlett et al., 2020).

Implications for Fetal Development

Having examined the social factors that may have influenced pregnant women's use of cannabis in Canada during the first years of the COVID-19 pandemic, I will now consider what this use means for fetal development. Cannabis use during pregnancy can have consequences for both maternal and fetal health outcomes (Agolli et al., 2022). Intrauterine life is understood to be a very sensitive time (Eleftheriades et al., 2022), during which fetal programming occurs, which is explained by Richardson et al. (2016) as "when internal or external maternal stressors experienced during pregnancy induce epigenetic changes in fetal somatic cells, which may lead to a specific postnatal, adolescence, and/or adult phenotype." Examples of internal stressors include mental health diseases, while external stressors include drug abuse and low socioeconomic status (2016). Thus, the impacts of the COVID-19 pandemic and prenatal cannabis use can be seen as stressors that influence fetal programming. During the intrauterine stage, health growth is largely dependent on maternal behaviours (Premji et al. 2019).

Potential Impacts of COVID-19 on Fetal Development & Infant Health

Not accounting for maternal cannabis use, babies born from pregnancies that happened during the COVID-19 pandemic may already have increased risks of negative health outcomes in life. This includes exposure to the SARS-CoV-2 virus during fetal development. A study on pregnancies in Canada completed between March 1, 2020, to October 31, 2021, found that "the rate of overall preterm birth (>37 weeks' gestation) increased in SARS-CoV-2-affected pregnancies, being 11.1% compared to 6.8 % among all unaffected Canadian pregnancies" (McClymont et al., 2022).

Preliminary research from a US study has found that infants born to mothers who had a positive SARS-CoV-2 test during pregnancy had an increased likelihood of receiving a neurodevelopmental diagnosis in the first 12 months after delivery, suggesting that COVID-19 exposure in-utero may be associated with neurodevelopmental changes (Edlow et al., 2022). It remains to be determined if COVID-19 can have long-lasting effects on neurodevelopment in childhood and adulthood (Volkow et al., 2021).

Increased rates of stress reported by pregnant women during the pandemic are also of concern. During pregnancy, the effects of stress may be transmitted from mother to child. Evidence suggests that exposure to excessive stress during fetal development can impact the development and health of infants in a variety of ways, both biologically and behaviorally. Maternal stress has been associated with chronic activation of the stress system, and may also be associated with unhealthy behaviours related to stress, including overeating and smoking (Eleftheriades et al., 2022).

Cannabis Composition

The two most well-known compounds in cannabis are THC (delta-9-tetrahydrocannabinol) and CBD (cannabidiol), which can impact health differently (Canadian Centre on Substance Abuse and Addiction). THC is the primary psychoactive ingredient in the plant and is the most well-known, while CBD or cannabidiol is the second ingredient. THC is generally viewed as the more concerning of the two in terms of use among the general population, while CBD is perceived as relatively safe. Though, CBD does have psychoactive effects upon exposure and can affect brain functions and behaviours (Grant et al., 2018). As previously mentioned, even if it may be the safer compound compared to THC in general cannabis consumption, the safety of CBD during pregnancy is unknown (Nashed et al., 2021).

Impacts on Fetal Development

Though research around the impacts of cannabis on fetal development and throughout life remains inconclusive with varying results, several factors have been repeatedly observed. Preterm birth and being small for gestational age have been identified as pregnancy outcomes associated with perinatal cannabis use. A literature review conducted by Baia et al. (2022) found that among 32 studies, cannabis use during pregnancy increased occurrences of low birth weight, preterm birth and being small for gestational age. Marchand et al. (2022) conducted a review aimed at assessing data on neonatal outcomes of in-utero cannabis exposure due to inconsistencies in study findings, analyzing 16 studies. It had findings similar to Baia et al., observing a significantly increased risk for preterm birth, low-birth weight and being small for gestational age. It also found an increased risk of NICU admission and a significant decrease in mean neonatal head circumference (2022).

After cannabis passes into the fetus's system through the placenta, it crosses the blood-brain barrier (Agolli et al., 2022). The structural remodelling the developing brain undergoes makes it "particularly vulnerable to harmful effects of bioactive ingredients" (Grant et al., 2018), such as those in cannabis. Though research has been inconsistent, possible influences of neurodevelopment that require further study include cognitive impacts on executive functioning skills and psychological health (2018).

Experiencing drug abstinence is very stressful for drug-dependent subjects. Offspring exposed to THC during fetal development have been observed to have increased drug-seeking behaviours, such as heroin, which could "reflect a behavioural response to stress, which intensifies the motivation for drug use" (Alpar et al., 2016). Data from the Maternal Health Practices and Child Development Study (MHPCD) which examined the effects of prenatal exposure to marijuana and alcohol on long-term development found increased use of marijuana in those exposed to it in-utero. The study initially recruited women in the fourth or fifth month of pregnancy and followed up with offspring at age 22. It was found that "offspring across all levels of PME had a higher percentage of marijuana initiation and a higher frequency of marijuana use compared to those without PME" (Sonon et al., 2015), with PME standing for prenatal marijuana exposure. Higher levels of prenatal exposure were also linked with an increased likelihood of marijuana use (2015).

Consumption

The mode of consumption of cannabis can influence the level of THC transferred to the fetus (Foeller et al., 2017). The rate of THC absorption can differ if oral or smoked. Thus, it is important to state that on October 17, 2019, cannabis edible products and concentrates became legal to sell in Canada (Department of Justice, 2021). Thus, by the start of the COVID-19 pandemic in the country, there was a broad array of cannabis products legally available for consumption.

THC rates and cannabis potency have significantly increased throughout the world in the past decade, which has implications for fetal cannabis use (Richardson et al., 2016). THC is rapidly absorbed with smoking, reaching peak concentrations within minutes, while oral consumption is considerably slower, with peak concentration reached within 1-3 hours after dosing. The peak concentration is slightly higher with smoking than with oral consumption when similar THC content is consumed. Route of consumption may impact overall fetal toxicity (Grant et al., 2018). Overalls factors of prenatal cannabis use that may impact the degree it influences fetal development, include potency (as influenced by THC levels), consumption levels, mode of consumption, and time of use during pregnancy.

Conclusion

Multiple stressful circumstances due to the COVID-19 pandemic impacted motivations for using cannabis while pregnant, including social isolation, increased rates of anxiety and depression, financial strife and food insecurity. Healthcare providers were challenged in giving adequate care and information regarding the dangers of cannabis use during pregnancy and its impacts on fetal development. This barrier was partly due to the recent of cannabis and healthcare providers' lack of knowledge on the harms of cannabis use while pregnant and how to counsel patients on the topic. In terms of the influence of prenatal cannabis use on fetal development, while much research remains to be done, the consensus among researchers is that cannabis use while pregnant is harmful to fetal development. It has been linked to outcomes growth outcomes such as low birth weight and placement in NICU (Baia et al., 2022; Marchand et al., 2022). It is also connected to neurodevelopmental outcomes and significantly increases the chance of marijuana use among offspring (Grant et al, 2018; Sonen et al., 2016).

The increase in average reported rates of cannabis use among pregnant women during the pandemic has drawn further attention from healthcare providers and other medical professionals to the need and potential for greater demand on the healthcare system. Though impacted by the willingness to disclose substance use after legalisation, this data suggests that among the demographic of babies born during the first years of the COVID-19 pandemic, from 2020-2022, there may be an increased risk for and higher rates of Fetal Cannabis Syndrome than is usual among the general population. This has implications for public infrastructure as increases in prenatal cannabis use could add further burden to the healthcare system in Canada (Koto et al. 2022, Ross et al. 2015).

This demographic may also face other developmental health problems due to other factors from the pandemic that can impact fetal health. Exposure to the SARS-CoV-2 virus in-utero has been associated with increased rates of preterm birth and may have possible impacts on neurological development (Edlow et al., 2022). Fetal health may also have been impacted due to exposure to high rates of maternal stress induced by the pandemic (Eleftheriades et al. 2022). Thus, healthcare professionals must be aware of the possible developmental impacts of the COVID-19 pandemic and prenatal cannabis on offspring born during this time in providing medical care, and the possible intersections of these two variables on development.

Though this chapter focused on the COVID-19 pandemic in Canada during 2020-2022, it is important to recognise that factors that may have influenced maternal health and cannabis use were not completely consistent throughout this time due to changing factors such as isolation restrictions. Maternal stressors, such as

mental health symptoms, and behaviours, such as levels of cannabis use are not static throughout pregnancy (Gesterling et al.,
2022; Vaghef-Mehrabani et al., 2022).

Moving forward, further longitudinal research is required on the
impacts of the COVID-19 pandemic on pregnant individuals
and how maternal cannabis use may influence fetal health outcomes in childhood and adulthood. Research studies on perinatal
cannabis use must isolate cannabis use from other drug use and
consider other common research limitations in their design, such
as self-reported use. It is also critical that physicians and other
health care providers be educated on perinatal cannabis use, understand its potential damage, and how to approach patients and
their questions.

References

Agolli, A., Agolli, O., Chowdhury, S., Shet, V., Canenguez Benitez, J.
S., Bheemisetty, N., & Waleed, M. S. (2022). Increased cannabis
use in pregnant women during COVID-19 pandemic. *Discoveries,
10(2)*. https://doi.org/10.15190/d.2022.7

Alpár, A., Di Marzo, V., & Harkany, T. (2016). At the tip of an iceberg:
Prenatal marijuana and its possible relation to neuropsychiatric
outcome in the offspring. *Biological Psychiatry, 79*(7), e33–e45.
https://doi.org/10.1016/j.biopsych.2015.09.009

Angus Reid Institute. (2020, April 27). *Worry, gratitude & boredom:
As COVID-19 affects mental, financial health, who fares better;
who is worse?* Retrieved February 19, 2023, from https://angusreid.
org/covid19-mental-health/

Bartlett, K., Kaarid, K., Gervais, N., Vu, N., Sharma, S., Patel,
T., & Shea, A. K. (2020). Pregnant Canadians' perceptions
about the transmission of cannabis in pregnancy and while
breastfeeding and the impact of information from health care
providers on discontinuation of use. *Journal of Obstetrics and
Gynaecology Canada, 42*(11), 1346–1350. https://doi.org/10.1016/j.
jogc.2020.04.015

Baía, I., & Domingues, R. (2022). The effects of cannabis use during
pregnancy on low birth weight and preterm birth: A systematic

review and meta-analysis. *American Journal of Perinatology.* https://doi.org/10.1055/a-1911-3326

Brown, R. A., Dakkak, H., Gilliland, J., & Seabrook, J. A. (2019). Predictors of drug use during pregnancy: The relative effects of socioeconomic demographic, and mental health risk factors. *Journal of Neonatal-Perinatal Medicine, 12*(2), 179–187. https://doi.org/10.3233/NPM-1814

Canadian Centre on Substance Use and Addiction. (n.d.). *Cannabis.* Retrieved March 1, 2023, from https://www.ccsa.ca/cannabis

Canada Department of Justice. (2021, July 7). *Cannabis legalisation and regulation.* Canada.ca. Government of Canada. Retrieved February 23, 2023, from https://www.justice.gc.ca/eng/cj-jp/cannabis/

Canadian Institute for Health Information. (2021). Unintended consequences of COVID-19: impact on harms caused by substance use. Ottawa, ON: CIHI.

Centre for Addiction and Mental Health. (2020, July). Mental Health in Canada: COVID-19 and beyond. CAMH.

Edlow, A. G., Castro, V. M., Shook, L. L., Kaimal, A. J., & Perlis, R. H. (2022). Neurodevelopmental outcomes at 1 year in infants of mothers who tested positive for SARS-CoV-2 during pregnancy. *JAMA Network Open, 5*(6), e2215787. https://doi.org/10.1001/jamanetworkopen.2022.15787.

Eleftheriades, M., Vousoura, E., Eleftheriades, A., Pervanidou, P., Zervas, I. M., Chrousos, G., Vlahos, N. F., & Sotiriadis, A. (2022). Physical health, media use, stress, and mental health in pregnant women during the COVID-19 pandemic. *Diagnostics, 12,* 1125. https://doi.org/10.3390/diagnostics12051125

Foeller, M. E., & Lyell, D. J. (2017). Marijuana use in pregnancy: Concerns in an evolving era. *Journal of Midwifery & Women's Health, 62*(3), 363–367. https://doi.org/10.1111/jmwh.12631.

Gesterling, L., & Bradford, H. (2021). Cannabis use in pregnancy: A state of the science review. *Journal of Midwifery & Women's Health, 67*(3), 305–313. https://doi.org/10.1111/jmwh.13293

Grant, K. S., Petroff, R., Isoherranen, N., Stella, N., & Burbacher, T. M. (2018). Cannabis use during pregnancy: Pharmacokinetics and effects on child development. *Pharmacology & Therapeutics, 182,* 133–151. https://doi.org/10.1016/j.pharmthera.2017.08.014

Health Canada. (2019, December 13). *Canadian Cannabis Survey 2019 - Summary.* Canada.ca. Government of Canada. Retrieved February 18, 2023, from https://www.canada.ca/en/health-canada/services/publications/drugs-health-products/canadian-cannabis-survey-2019-summary.html

Health Canada. (2021, December 08). *Canadian Cannabis Survey 2020.* Canada.ca. Government of Canada. Retrieved February 14, 2023, from https://www.canada.ca/en/health-canada/services/drugs-medication/cannabis/research-data/canadian-cannabis-survey-2020-summary.html

Holland, C. L., Nkumsah, M. A., Morrison, P., Tarr, J. A., Rubio, D., Rodriguez, K. L., Kraemer, K. L., Day, N., Arnold, R. M., & Chang, J. C. (2016). "Anything above marijuana takes priority": Obstetric providers' attitudes and counseling strategies regarding perinatal marijuana use. *Patient Education and Counseling, 99,* 1446–1451. https://doi.org/10.1016/j.pec.2016.06.003

Hwang, T. J., Rabheru, K., Peisah, C., Reichman, W., & Ikeda, M. (2020). Loneliness and social isolation during the COVID-19 pandemic. *International Psychogeriatrics, 32*(10), 1217–1220. https://doi.org/10.1017/S1041610220000988

Kar, P., Tomfohr-Madsen, L., Giesbrecht, G., Bagshawe, M., & Lebel, C. (2021). Alcohol and substance use in pregnancy during the COVID-19 pandemic. *Drug and Alcohol Dependence, 225,*108760. https://doi.org/10.1016/j.drugalcdep.2021.108760

Koto, P., Allen, V. M., Fahey, J., & Kuhle, S. (2022). Maternal cannabis use during pregnancy and maternal and neonatal outcomes: A retrospective cohort study. *BJOG, 129,* 1687–1694. https://doi.org/10.1111/1471-0528.17114

Marchand, G., Masoud, A. T., Govindan, M., Ware, K., King, A., Ruther, S., Brazil, G., Ulibarri, H., Parise, J., Arroyo, A., Coriell, C., Goetz, S., Karrys, A., & Sainz, K. (2022). Birth outcomes of neonates exposed to marijuana in utero. *JAMA Network Open, 5*(1), e2145653. https://doi.org/10.1001/jamanetworkopen.2021.45653

McClymont, E., Albert, A., Alton, G., Boucoiran, I., Castillo, E., Fell, D. B., Kuret, V., Poliquin, V., Reeve, T., Scott, H., Sprague, A., Carson, G., Cassell, K., Crane, J., Elwood, C., Joynt, C., Murphy, P., Murphy-Kaulbeck, L., Saunders, S., ... Money, D. (2022). Association of SARS-CoV-2 infection during pregnancy with maternal and perinatal outcomes. *JAMA Network*, 1983–1991. https://doi.org/0.1001/jama.2022.5906.

Nashed, M. G., Hardy, D. B., & Laviolette, S. R. (2021). Prenatal cannabinoid exposure: Emerging evidence of physiological and neuropsychiatric abnormalities. *Frontiers in Psychiatry*, *11*, Article 624275. https://doi.org/10.3389/fpsyt.2020.624275

Premji S., McDonald S.W., Zaychkowsky C., & Zwicker J.D. (2019). Supporting healthy pregnancies: Examining variations in nutrition, weight management and substance abuse advice provision by prenatal care providers in Alberta, Canada. A study using the All Our Families cohort. *PLoSONE 14***,** e0210290. https://doi.org/10.1371/journal.pone.0210290

Richardson, K. A., Hester, A. K., & McLemore, G. L. (2016). Prenatal cannabis exposure - The "first hit" to the endocannabinoid system. *Neurotoxicology and Teratology*, *58*, 5–14. https://doi.org/10.1016/j.ntt.2016.08.003

Ross, E. J., Graham, D. L., Money, K. M., & Stanwood, G. D. (2015). Developmental consequences of fetal exposure to drugs: What we know and what we still must learn. *Neuropsychopharmacology*, *40*, 61–87. https://doi.org/10.1038/npp.2014.147

Rotermann, M. (April 2021). Looking back from 2020, how cannabis use and related behaviours changed in Canada. *Health Reports, 23*(4). https://doi.org/10.25318/82-003-x202100400001-eng

Shirreff, L., Zhang, D., DeSouza, L., Hollingsworth, J., Shah, N., & Shah, R. R. (2021). Prevalence of food insecurity among pregnant women: A Canadian study in a large urban setting. *Journal of Obstetrics and Gynaecology Canada, 43*(5), 1260–1266. https://doi.org/10.1016/j.jogc.2021.03.016

Sonon, K., Richardson, G., Cornelius, J. R., Kim, K. H., & Day, N. L. (2015). Prenatal marijuana exposure predicts marijuana use in offspring during young adulthood. *Neurotoxicology and Teratology*, *47*, 10–15. https://doi.org/10.1016/j.ntt.2014.11.003

Vaghef-Mehrabani, E., Wang, Y., Zinman, J., Beharaj, G., van de Wouw, M., Lebel, C., Tomfohr-Madsen, L., & Giesbrecht, G. F.

(2022). Dietary changes among pregnant individuals compared to pre-pandemic: A cross-sectional analysis of the pregnancy during the covid-19 pandemic (PDP) study. *Frontiers in Nutrition, 9,* 997236. https://doi.org/10.3389/fnut.2022.997236

Vanstone, M., Panday, J., Popoola, A., Taneja, S., Greyson, D., & Darling, E. (2022). Pregnant people's perspectives on cannabis use during pregnancy: A systematic review and integrative mixed-methods research synthesis. *Journal of Midwifery & Women's Health, 67*(3), 354–372. https://doi.org/10.1111/jmwh.13363

Volkow, N. D., Gordon, J. A., & Freund, M. P. (2021). The healthy brain and child development study—Shedding light on opioid exposure, covid-19, and health disparities. *JAMA Psychiatry, 78*(5), 471–472. https://doi.org/10.1001/jamapsychiatry.2020.3803

Afterword

Holly Steen

This book has examined topics related to Fetal Cannabis Syndrome from both biological and social perspectives. It has discussed various behavioural, cognitive and physical impacts on biological development that in-utero exposure to cannabis can have. This includes impacts on the fetal brain and neurological development that can be influential throughout life, impacting cognitive function, and intelligence rates. Cannabis can impact other parts of the body beyond the brain, including the immune system and immune response. Biological impacts may vary according to factors such as the time of use during pregnancy, amounts consumed, and potency. Factors that may influence the choice to consume cannabis while pregnant include being younger, lower income, having an addiction, suffering from mental health issues and a lack of education (both in general and regarding the dangers of cannabis use).

As cannabis becomes an increasingly accepted substance within society and is legalised in more countries throughout the world, further definitive research is required on prenatal cannabis use, its impact on maternal health, fetal development and long-term development in life. The influences of legalisation and the COVID-19 pandemic on prenatal cannabis use are in the early stages of research and remain to be seen. Further research may aid in educating healthcare providers on counselling pregnant women on cannabis use, providing treatment, and be beneficial for creating public campaigns aimed at awareness.

www.ingramcontent.com/pod-product-compliance
Lightning Source LLC
Chambersburg PA
CBHW030853270326
41928CB00008B/1347